Overseas R&D Activities of Transnational Companies

Overseas R&D Activities of Transnational Companies

Jack N. Behrman
University of North Carolina

William A. Fischer
University of North Carolina

Sponsored and administered
by the

**Fund for Multinational
Management Education**

Oelgeschlager, Gunn & Hain, Publishers, Inc.
Cambridge, Massachusetts

International Standard Book Number: 0-89946-016-x

Library of Congress Catalog Card Number: 79-25296

Printed in the United States of America

Library of Congress Cataloging in Publication Data

Behrman, Jack N
    Overseas R&D activities of transnational companies.

    Includes bibliographical references and index.
    1. Research, Industrial.   2. International business enterprises.   I. Fischer, William A., joint author. II. Title.
HD30.4.B44   338.4'7'6072   79-25296
ISBN 0-89946-016-x

The views presented in this book do not necessarily represent those of the Directors, Officers, or Staff of the Fund for Multinational Management Education.

Contents

List of Figures

List of Tables

Preface

This study of the overseas R&D activities of transnational companies is an outgrowth of the authors' long involvement in problems of technology transfer and technology management. It became clear several years ago that governments were becoming increasingly concerned with not just technology transfer but also with technology generation: the advanced countries with their ability to remain at the leading edge of scientific advance, and the developing countries with their ability to establish the foundations of scientific and technical activity.

The transnational companies are, of course, the principal generators of technology in the private sector, being able to take advantage of the international markets for their particular expertise. Consequently, their role in the advanced and developing countries is certain to become the focus of policy dialogues and agreements among governments.

It has been the objective of the several studies encompassed in this volume to provide a better understanding of the process of undertaking R&D across national boundaries by the transnational corporations so that policy dialogues would be better founded on an appreciation of the patterns employed and problems faced by these companies.

Because of the paucity of detailed information on company activities, the senior author set out to learn the entire process of R&D within transnationals. The order was, of course, too large; it had to be cut into digestible pieces. A number of companies were selected to help formulate the project and even finance some of the costs. After some months of discussion, five companies had agreed to participate. The case studies were done sequentially over a two-year period, with interviews at headquarters laboratories and some of the overseas laboratories as well. The inability to encompass the entire R&D activities of a company required that a certain product group, or product line, or set of activities, be taken as the focus. This limitation was not serious, for the purpose was not to describe one company fully, but to learn enough of the entire process to be able to take the next steps in formulating a more generalized study.

Each of the completed case studies were submitted to the company involved for checking as to accuracy and for removal of confidential materials, since officials were permitted by headquarters to speak without any restraint during the interviews. In all but one company, the interviewer was accompanied by a senior R&D management official for the purposes of assuring the respondents that they could be candid and of helping to translate the questions posed by the interviewer so that responses would be comprehensive and useful. It was a process of mutual learning, and headquarters themselves discovered some aspects of R&D operations that were not known or appreciated before.

In several instances, the knowledge gained helped in the reformulation of R&D objectives and organization. Consequently, in none of the three case studies included in this volume is the R&D activity currently like that represented here. Considerable discussion was engendered over this anomaly, but in each case, the company determined to let history be told, rather than trying to update their story—the principal reason being that even that data would be out of date by the time of publication. The cases are, therefore, snapshots of a past time period and do not show the dynamism and constant change within the R&D function. They must be read as examples of R&D activities—*not* as the way in which Unilever, Johnson & Johnson, or Du Pont presently carries out its research activity, even within the product lines described.

Although the case studies were done preparatory to an effort to generalize their results, as recorded in the first chapters of this volume, they are placed at the end, to serve as detailed illustrations of the generalized process. These case studies were done by the senior author prior to the junior author joining the project. The joint effort began with the multiple interviews of some fifty-six companies, with Professor Fischer doing virtually all of the interviews with American companies, and Professor Behrman conducting those with companies in Europe and Japan. Both sets of interviews were conducted during the summer of 1978, so the basic materials underlying this section are more current than those of the case studies. The writing of the results of each interview was the responsibility of each interviewer, in turn, but Professor Fischer was responsible for the synthesizing and writing of the chapters recording the results of these multiple interviews.

It is obvious that we could not have conducted this study without the considerable help of a number of company officials. Our debt to them is great, particularly for their high level of confidence in our ability and willingness to report accurately and faithfully what was seen and heard. In addition, the unstinting time given by officials not only in interviews but in accompanied trips and in reviewing manuscripts must be considered as an contribution of immeasurable value. We can only thank them in this wholly inadequate way and hope that the results of the study will be worth the time committed, both in producing more informed dialogues on the future role of TNCs in R&D and in policies on science and technology.

Although we promised that no company officials would be mentioned among those we interviewed and that no attribution would made, save as permitted in the final review of case manuscripts, we cannot avoid mentioning the encouragement and assistance given by Dr. Bruce Hannay, Foreign Secretary of the National Academy of Engineering and Director of Bell Labs, in formulating the study and selecting the companies for study. His recommendation to the Industrial Research Institute was instrumental in our obtaining their cooperation in the selection of companies and in their assistance in undertaking the project. This assistance extended also to obtaining cooperation from the European and Japanese companies as well.

Finally, we must thank the officers of the National Science Foundation who considered various aspects of our larger project worth funding and the Fund for Multinational Management Education of New York, which supported and administered the research project.

Jack N. Behrman
William A. Fischer
Chapel Hill, N.C.

Introduction

During the past few years concern over diminished American industrial competitiveness, creation of appropriate technology for developing countries, technology transfer processes, and the formation of a new international economic order have raised considerable interest in the conduct of research and development (R&D) by the foreign affiliates of transnational corporations. Although interest is high, however, little information is available. What few studies do exist have mostly focused on aggregate statistics without descriptions of activities or underlying motivations and influences. The result is that we know very little about what the statistics purport to tell us. What type of R&D is performed by these companies in foreign sites, for example, and how have they come to select these sites and assign various projects to them? How do transnational corporations manage R&D performed in foreign locations, and does this affect patterns of invention and innovation? What sort of collaborative R&D activities are the transnational corporations engaged in? What contribution does the R&D performed by transnational corporations in foreign locations make to relevant host-country communities? What is the nature of the relationship between host-country governments and transnational corporations regarding the performance of R&D in foreign locations?

THE MOTIVATION
FOR THIS STUDY

The answers to these questions are of particular interest to three different groups of policymakers: those in transnational corporations, faced with decisions concerning the placement of R&D abroad and desirous of learning what others have done; those in developing countries, faced with a need for improving their scientific and technical capabilities and tending to regard transnational corporations as both sources of technology and potential generators of local R&D capabilities; and those in the parent nations of the transnational corporations, who desire to transfer technology to developing countries, yet who must understand what the extent and consequences of such technology (including R&D) transfer will be both for their own nation and for the developing country. It is to these three groups of policymakers that this study is addressed.

The motivation to undertake this study grew out of the increasing importance of the questions associated with the issue of foreign R&D activities by transnational corporations, and the paucity of information available to answer these questions. While some attention has been paid to the transnationals' performance of R&D abroad, for the most part the utility of these studies to the policymaker has been limited by the small number of observations they have drawn from[1], or, conversely, by statistical aggregation (typically financial) so large that it prevents a real understanding of the underlying decision structure.[2] The present research originated with five detailed case studies, three of which are included in this study. In the course of completing these case studies a number of findings arose that could not be examined for generality because of the limitations of the existing literature. It was then decided to examine these issues in a less detailed form than the case studies, but with a larger sample of observations, in order to ascertain their presence outside of the original sample.

STUDY DESIGN AND OBJECTIVES

The process of generalization led to the selection of specific topics to be examined and the companies to be interviewed—neither of which was done randomly. Six major topics were chosen: location abroad, R&D management, market orientations, international collaboration, diffusion, and government policies in host countries.

The motivation to do R&D abroad, the choice of foreign locations, the assignment of missions to foreign R&D groups, and the means of establishing R&D groups abroad, are discussed in Chapter 1. The objectives of this chapter are to determine:

The patterns of foreign R&D activities of the transnational firms in our sample.

How American transnationals compare with their European counterparts in the performance of R&D abroad.

The status of transnational corporations' R&D in developing and developed countries.

Whether the market-orientations of the transnational corporations affects the missions and locations of their foreign R&D activities.

How the transnationals establish R&D abroad and how it is affected by the foreign laboratory's mission and the parent firm's market- and science-orientation.

The management styles exhibited in the transnational corporations and their effect on the coordination and control of foreign R&D activities are examined in Chapter 2. In it we record:

How the multinational coordination of R&D is affected by the firm's market- or science-orientation.

What significant differences exist between European and American transnationals in R&D management styles and coordination and control systems.

The ultimate purpose of doing industrial R&D is to foster invention and innovation. Accordingly, chapter three examines how market-orientation and management styles affect the patterns of invention and innovation in the foreign laboratories of transnational corporations. Additionally, R&D group size and international product and process standardization are examined, since both are closely interrelated with invention and innovation.

R&D can be transferred between entities of differing nationalities in a variety of ways. Direct foreign investment in R&D facilities, as described in the first three chapters, is just one of the ways. Chapter 4 examines several collaborative international R&D activities, including joint manufacturing ventures requiring R&D support, joint R&D ventures, and cooperative R&D projects. This chapter describes:

The patterns of international collaborative R&D activities.

The ownership position of the partners in such collaborative arrangements and its effect on participation.

The motivations leading to such collaborative activities.

The pursuit of R&D by transnational corporations in a foreign country is often seen by the host-country government as a vehicle for transferring R&D related skills and capabilities to the host-country's scientific and technical communities. If this is, in fact, the case, then it makes the establishment of such activities especially attractive to developing country governments, and it provides the transnational corporation with additional bargaining points in gaining access to a market, or favorable investment conditions. Chapter 5 characterizes:

The diffusion of R&D related skills and capabilities to the scientific and technical communities of the host country.

The effectiveness of transnational corporations' employing local nationals as R&D professionals or as R&D managers.

The use of R&D as a training ground for host-country nationals, the upgrading of supplier and customer capabilities, and the provision of technical and scientific assistance to host-country universities and institutions as vehicles for diffusing R&D related skills and capabilities to host-country scientific and technical communities.

The nature of planning by the transnational corporations for the diffusion of R&D related skills and capabilities.

Of particular importance to each of the three groups of policy-makers addressed by this study is the nature of the relationships between host-country governments and the transnational corporations regarding the location and performance of R&D activities. Chapter 6 discusses:

The pressures transnational corporations have felt from foreign governments to locate R&D abroad.

The forms this pressure has taken and, the effectiveness of this pressure.

FIRMS SURVEYED

The information presented in this report was gained from a series of structured interviews conducted at the headquarters of thirty-four transnational corporations in the United States, sixteen in Europe, and six in Japan, during the spring and summer of 1978. The firms interviewed were not randomly selected but, rather, reflected some deliberate situations or experience judged significant by the authors and by knowledgeable observers of international industrial R&D. Since a number of recent examinations of R&D investment patterns have concluded that not much R&D is being performed abroad by transnational corporations regardless of nationality,[3] it was decided to concentrate on firms known to have some foreign R&D in order to yield a useful set of experiences. Accordingly, the sample of American firms was selected from the membership of the Industrial Research Institute (IRI), which comprises firms with substantial domestic research commitments; and who we had reason to believe had foreign R&D activities or experiences. A second, small, group of American firms was also chosen from among Industrial Research Institute members because we knew that they had no foreign R&D activities of their own. These few firms had high international commercial exposure, and it was considered useful to understand why such firms were not involved in foreign R&D activities. (One additional factor considered in the selection of a sample of American firms was a limitation to firms on the east coast and in the midwest to reduce travel expenses).

The selection of European and Japanese firms reflected "targets of opportunity" to the extent that the firms were prominent and research intensive, and the interviewer had prior contacts with them that facilitated entreé for this study. These prior contacts were especially important to this study because of the sensitivity of the subject of R&D activities to the firms involved.

NOTES

1. Among the works in this category are: W. T. Hanson, "Multinational R&D in Practice. Eastman Kodak Corporation," *Research Management*, Jan. 1971, pp. 47-50; M. Papo, "How to Establish and Operate a Multinational Lab," *Research Management*, Jan. 1971, pp. 12-19; R. Van Rumker, "Multinational R&D in Practice: Chemagro Corporation," *Research Management*, Jan. 1971, pp. 5054; U.S. National Academy of Sciences, *U.S. International Firms and R.D.&E. in Developing Countries*, (Washington, D.C., National Academy of Sciences, 1975); L.W. Steele, *Innovation in Big Business*, (New York, Elsevier, 1975); V. Terpstra, "International Product Policy: The Role of Foreign R&D," *The Columbia Journal of World Business*, Vol. 12, No. 4, Winter 1977, pp. 24-32; R. C. Ronstadt, "International R&D: The Establishment and Evolution of Research and Development Abroad by Seven U.S. Multinationals," *Journal of International Business Studies*, Vol. 9, No. 1, Spring/Summer 1978, pp. 7-24.
2. Examples of this group of work include: D. Creamer, *Overseas Research and Development by United States Multinationals, 1966-1975*, (New York, The Conference Board, 1976); Industry Studies Group, Division of Science Resources Studies, U.S. National Science Foundation, "U.S. Industrial

R&D Spending Abroad," *Reviews of Data on Science Resources*, NSF 79-304, No. 33, April 1979.

3. Several recent studies of transnational corporations all indicate a low level of foreign R&D activities by these firms. In the case of American firms, it was reported that in 1966 only 537 U.S. companies had foreign R&D expenditures and that, in the early 1970s, foreign R&D expenditures amounted to about 10 percent of domestic corporate R&D expenditures by all U.S. firms. [D. Creamer, ibid., pp. 3-4]. By 1977, U.S. transnational corporate spending on R&D abroad was estimated to be $1.5 billion, yet it was still noted that "The vast majority of U.S. firms do not conduct any research and development abroad. According to the Census Bureau, only an estimated 15 percent of the major U.S. industrial R&D performers maintain foreign laboratories. These firms, however, account for nearly one-half of total U.S. company-funded R&D expenditures. It is highly unusual for smaller U.S. firms to have a foreign R&D facility. . . .Foreign R&D spending by U.S. firms . . . amounts to about 7 percent of industry's company R&D funds. "[Industry Studies Group, ibid., p. 3]. Franko has concluded that "It [is] unusual for a Continental European firm to allow any foreign subsidiary to do its own R&D," but adds that "currently available data tell one little about the propensity of European firms to undertake R&D activities outside of their home-country markets. No systematic data seems available on American practice either. The general rule in both American and European practice seems to have been to keep R&D at home." [Lawrence G. Franko, *The European Multinationals*, (Stanford, Connecticut, Greylock Publishers, 1976) pp. 181]. Japanese firms appear even less likely than American or European transnationals to entertain thoughts of foreign R&D activities. Parker has concluded that "apparently Japanese companies are independent of the locational influence which induce other concerns in the population [of research-intensive multinational corporations] to locate abroad. Classification by multinational status and research intensity, reveals a pattern which is seemingly unlike that on the rest of the world." [J. E. S. Parker, *The Economics of Innovation*, (London, Longman, 1974), p. 165.]

PART I

Overseas R&D Activities of Transnational Companies: A General Perspective

Chapter 1

Locations, Missions, Motivations, and Growth Paths of Foreign R&D Activities

An examination of the case studies performed as part of this research and a reading of the small amount of available literature suggest that there is little R&D conducted abroad by American companies, less by Europeans, and practically none by the Japanese. There is, however, precious little data to support such observations, other than on the most general grounds. Additionally, there has been no concerted effort, on a large, multinational basis, to examine the relationships among R&D activity abroad, location of foreign laboratories, missions of foreign laboratories, means of establishing R&D abroad, and the motivations of the transnational corporations. From our interviews, we are able to shed a bit more light on these aspects.

THE LEVEL AND LOCATION OF R&D ACTIVITY ABROAD

One would assume, from the literature and from casual conversations with individual R&D managers about the level and location of R&D activities abroad, that not much is being done. Yet, within the thirty-one (out of thirty-four) American transnational firms that had any R&D activities abroad there were 106 active foreign R&D groups. This amounts to slightly more than three foreign R&D activities per firm.

The interviews with sixteen European firms identified 100 distinct foreign R&D activities, averaging more than five per firm. The six Japanese firms reported no foreign R&D activities, and none of the officials interviewed could think of a company that had or was likely to have an R&D unit overseas at present.

In all cases, only activities that were clearly research and development were included. Technical service and quality control activities were not included in these figures because of the difficulty of ascertaining their R&D content and because of the exceedingly wide diversity of these operations even within the companies interviewed.

Over thirty different countries were identified in our interviews as hosting the foreign R&D activities of the firms in the sample. Both American and European firms indicated that their foreign R&D activities were located predominantly in developed countries such as the United States, France, the United Kingdom, Japan, Canada, Australia, and Germany, and in advanced developing countries such as Mexico, Brazil, and India (Tables 1.1 and 1.2. Since Japanese firms interviewed have no foreign R&D activities, they are not represented in these tables).

THE MISSIONS OF FOREIGN R&D ACTIVITIES

Simply counting up and comparing numbers of foreign R&D activities reveals very little of the nature of the R&D being pursued abroad. To determine what actually is occurring in the foreign R&D activities of transnational corporations requires an assessment of the missions of the laboratories. Here, the conventional wisdom is that R&D performed abroad is typically development-oriented in the late stages of innovation.

In the 106 foreign R&D groups belonging to American transnational corporations, approximately 28 percent (thirty foreign R&D groups) had missions that included a substantial commitment to new product research.[1] Fifty-nine percent (sixty-two foreign R&D activities) had missions limited to applied research and development. Table 1.3 provides a complete breakdown of the sample of activities considered. Of the thirty R&D groups that performed new product research, at least four of them (two in Switzerland, one in Japan, and one in Germany) were expressly concerned with only exploratory, or basic, research.

Table 1.1. Country of Location of Foreign R&D Activities of Thirty-one American Transnational Corporations (Observations Limited to American Firms and Multiple Activities Within a Nation Counted as One)

Country	Number of American Firms with R&D Activities Located in Each Country
United Kingdom	11
Australia	8
Canada	8
Japan	8
France	7
Germany	6
Mexico	6
Brazil	5
Netherlands	4
Switzerland	3
Hong Kong	2
India	2
Italy	2
Spain	2
Argentina	1
Austria	1
Belgium	1
Columbia	1
Denmark	1
Ecuador	1
Egypt	1
New Zealand	1
Norway	1
Philippines	1
South Africa	1
Sweden	1
Taiwan	1

Source: Interviews

Out of the 106 foreign R&D activities of the U.S. firms, eighty-eight were specifically identified as to both location and mission. Those activities that had a substantial new product research commit-

Table 1.2. Country of Location of Foreign R&D Activities of European Firms (Multiple Activities Within A Nation Counted as One)

Country	Number of European Firms With R&D Activities Located in Each Country
United States	14
France	10
Germany	9
India	6
Brazil	5
United Kingdom	5
Italy	4
Japan	4
Spain	4
Sweden	4
Argentina	2
Austria	2
Belgium	1
Canada	1
Greece	1
Netherlands	1
Hong Kong	1
Mexico	1
Singapore	1
South Africa	1
Switzerland	1

Source: Interviews.

ment were almost entirely located in Western Europe. The prominence of Western European locations was again evident, although to a lesser degree, when those activities dedicated to applied research and development were considered. The breakout of laboratory location and mission for the American firms is shown in Table 1.4. (As noted earlier, in many, if not most of the firms interviewed, strict technical service activities were not considered to be a part of R&D and thus were not covered in the interviews. In those few instances where they were discussed, the data were not included in

Table 1.3 Missions of the Foreign R&D Activities of the Thirty-one American Transnational Corporations

Mission includes a substantial amount of new product research	30
Mission limited to applied research and development	62
Animal and farm facilities	9
Regional scientific and clinical staffs	5
	106

Source: Interviews.

Tables 1.3 or 1.4. It was, however, evident that the geographic distribution of technical services is much broader than that of research activities, and includes considerably more developing countries.) As might be expected, the location of animal and farm facilities for testing of veterinary products, pesticides, and agricultural products was largely in tropical, developing countries.

The European transnationals, like their American counterparts, have a preference for advanced industrialized nations as locations for foreign R&D activities. However, they differ in two important respects. First, the European firms appear to be considerably more willing to assign new product development responsibility to foreign R&D activities than do American firms. Given the close physical proximity of European locations, however, and the historical corporate alliances existing among the European firms, this finding may be more reflective of geographic and cultural proximity than of corporate behavior. Second, European firms are apparently more willing to pursue new product research in the advanced developing (or newly industrialized) countries.[2]

The absence of overseas R&D activities by Japanese firms is a result of their historical development and mission as they presently conceive it. This was explained by one Japanese manager as follows: "Japanese companies have been importing technology, and their R&D laboratories have been responsive to the needs of these transfers and adapting and digesting foreign technology rather than creating their own. As the Japanese begin to develop their own technology they will also begin to originate more research and then sell it to others. To date, little Japanese technology has been developed, and it will probably require some ten years to do so. Establishment

Table 1.4. Frequency of Location, by R&D Categorization, for Overseas R&D Activities by American Firms

Mission Includes New Product Research		Mission Limited to Applied Research and Development		Regional Scientific Staff		Animal and Farm Facilities	
Switzerland	2	Canada	4	Hong Kong	1	Netherlands	1
Japan	3	Norway	1	Switzerland	1	Brazil	2
Canada	3	France	4	Brazil	1	South Africa	1
France	5	Brazil	3	Italy	1	Taiwan	1
Spain	1	United Kingdom	7	Mexico	1	Egypt	1
Germany	4	Australia	6	Total	5	Philippines	1
United Kingdom	6	Germany	3			Spain	1
Netherlands	2	Netherlands	1			United Kingdom	1
Australia	1	Mexico	4			Total	9
Austria	1	Japan	4				
Italy	1	Argentina	1				
Sweden	1	Colombia	1				
Total	30	Ecuador	1				
		New Zealand	1				
		Denmark	1				
		India	1				
		Italy	1				
		Total	44				
		One firm whose facilities not explicitly identified	18				

Source: Interviews.

14

of overseas R&D laboratories will come after that." In addition, the Japanese style of centralizing control of all affiliates, including those overseas, forces R&D into a centralized activity, not for secrecy's sake, but to reduce uncertainty. This inhibits the initiation of R&D activities overseas. Furthermore, some Japanese managers feel that a lack of sophistication on the part of Japanese scientists and engineers prevents their adequately supporting the establishment of overseas laboratories.

MARKET ORIENTATIONS, R&D MISSIONS, AND FACTORS CONSIDERED IN THE LOCATION DECISION

In order to understand why a firm pursues R&D abroad, assigns a mission to a foreign laboratory, or selects a particular location, it is necessary to know the firm's orientation to international markets. The firms interviewed were categorized into three different market orientations.

The first group consists of those companies whose primary market orientation is to their home market. The companies in this "home-market" group are typified by the extractive industries and offshore component manufacturers (involved in what is sometimes called "foreign sourcing"). They can be expected to have few enduring overseas scientific and technical commitments, a highly ethnocentric managerial style, low external orientation on the part of management, and an organizational structure that is unaffected by the firm's foreign operations. By and large, the products of these firms are highly standardized, often because of their commodity nature. In this study there are seven such companies, all American, largely in the nonferrous metals and commodity chemical industries.

The second category includes those firms whose primary orientation is to the particular national markets where they are located. These companies are designated "host-market" firms and include industries spanning a range of technical sophistication from chemicals and pharmaceuticals to foods and tobacco. The products of the firms in this group typically exhibit a high degree of standardization within a market but not necessarily between markets. The management style of these companies can be characterized as polycentric, since a relatively low level of control is exerted by corporate headquarters over a decentralized organizational structure. There are twenty-three such firms among the American companies in our sample—seven pharmaceutical firms, four chemical firms, and four food and tobacco firms—plus fifteen European firms.

The third category comprises those companies with an international market orientation. The firms in this "world-market" group are typically characterized by a high degree of worldwide product standardization, a highly centralized corporate structure, a geocentric (one management for the world) management style, and high technology. In our sample, four of the five "world-market" firms interviewed are in electronics-related industries. All but one of the "world-market" companies are American.

Market Orientations and Level of R&D Activities and Missions

The literature on the foreign R&D activities of transnational corporations suggests that the market orientations of these firms should affect their R&D location and mission decisions. As expected, the large majority of the foreign R&D activities identified among the American transnational companies belongs to firms with a host-market orientation (96 out of the 106 foreign R&D activities identified, or 91 percent). While this reflects, to some extent, the preponderance of host-market companies in the sample (twenty-three of the thirty-four—68 percent—American firms had host-market orientations), it goes well beyond a simple proportional relationship. The host-market companies average more than four foreign R&D activities per firm, compared with less than two among the four American world-market companies, and slightly less than one-half a foreign R&D activity for each of the seven American home-market companies.

The prior discussion on the location of foreign R&D activities and their missions indicated that 30 of the 106 (28 percent) American transnational corporations' foreign R&D activities had missions that included new product research (Table 1.4). Of these thirty facilities, twenty-five belonged to host-market companies and five belonged to world-market companies. These data suggest that world-market firms are more likely to assign new product research to their foreign R&D groups than are firms with other market orientations, as "new product" world-market laboratories represent 71 percent of the total foreign R&D laboratories belonging to world-market companies while the twenty-five host-market laboratories represent 26 percent of the host-market companies' foreign R&D laboratories. None of the home-market companies had new product research missions (Table 1.5 and 1.6) Furthermore, three of the five foreign lab-

Table 1.5. Market Orientation and Inclusion of New Product Research in the Mission of Foreign Laboratories

Market Orientation	Number of Laboratories with a New Product Research Mission	Percentage of Foreign Laboratories with a New Product Research Mission
"World-market"	5	71%
"Host-market"	25	26%
"Home-market"	0	0%

Table 1.6. Market Orientation and Foreign R&D Laboratories with an Exploratory Research Mission

Market Orientation	Number of Laboratories with an Exploratory Research Mission	Percentage of Foreign Laboratories with an Exploratory Research Mission
"World-market"	3	60%
"Host-market"	2	8%
"Home-market"	0	0%

oratories dedicated to exploratory research belonged to world-market firms.

Market Orientations and Other Factors Considered in the Foreign R&D Location Decision

There are both strong similarities and strong differences among the factors considered by firms with different market orientations in the foreign R&D location decision. In general, the most powerful inducements to locate R&D in a particular foreign location appear to be the presence of a profitable affiliate in the foreign country, and a growing and sophisticated market with an adequate scientific and technical infrastructure. The primary obstacles to locating R&D abroad appear to be the firms' perceptions of the economies of centralized R&D and the perceived difficulties of assembling an adequate R&D staff in foreign countries. There was virtually unanimous agreement among managers interviewed that the days of doing R&D abroad to save money are gone; one executive, with a firm having nine foreign R&D groups, observed that only in Colombia was it actually cheaper to do R&D than in the United States.

Home-Market Companies. Those companies oriented to their home market were primarily concerned with tying foreign R&D activities directly to their foreign manufacturing and extraction operations. For the most part, the extractive firms do not do much processing or fabrication abroad. Their foreign commercial activities are chiefly limited to getting their raw materials out of the country of origin. Similarly, component manufacturers (in electronics or textiles, for example) have placed manufacturing units abroad to take advantage of the low wages associated with labor-intensive assembly operations. To the limited extent that these firms actually market products outside of their home market, they are concerned with merely providing exports of products developed for and sold primarily in the home market. Accordingly, any motivation they might have for performing R&D abroad is largely related to the technical support of their foreign extractive or manufacturing activities; typically they locate their foreign R&D activities at a mine site or manufacturing facility. This close attachment to operations allows the R&D facility to draw upon the operation's resources for support but at the same time limits the scope of the research activities abroad to a relatively narrow process focus.

A second, related, criterion for overseas R&D activity location among home-market firms appears to be the availability of adequate local universities. The firms in the home-market category tend to employ a large number of engineers and technicians in the performance of their day-to-day extractive operations. The availability of such individuals and, perhaps more important, the availability of analytical testing and even small pilot plant facilities typically found at universities, is a key consideration in the foreign R&D location decision. The motivation here is to employ adequately trained local technicians to perform rather low grade technical tasks. One of the firms in this category had a substantial R&D effort in one Latin American nation but none at all in the other Latin American nations it operated in, solely because of the availability of pilot plant facilities and competent faculty and graduate students at a local university at the one site. The company believed this combination to be unusual in a developing country, and it took advantage of the situation. The executive interviewed felt that universities took on added importance in the developing countries because, in the absence of engineering firms, the universities were the most important repository of technical skills.

Host-Market Companies. Firms in the host-market group locate R&D abroad to address the diverse needs of local markets. For example, an agricultural chemical firm cited the need to test its products in their intended markets and thus has facilities in South America to treat South American pest problems, facilities in the Far East to address problems of tropical climates, and facilities in the Philippines that provide market conditions indicative of Japan's. A number of pharmaceutical companies reported a need for different drug formulations for different markets; the European preference for injectibles and suppositories, compared with American preferences for orally administered formulations, illustrate why all of the American pharmaceutical firms interviewed have European formulation laboratories. An energy-based firm cited the unique problems of Canada's cold-weather environment as the principal reason for its locating R&D efforts there. Several tobacco and food firms cited not only differing national preferences for their products but also different raw materials existing in various markets, which required R&D efforts. In sum, the need to be close to their demand and raw materials are the most important motivating influences in the host-market firm's foreign R&D location decision. But these are necessary, not sufficient, conditions for establishing R&D abroad.

As market size increases, the potential ability to support R&D resources in a foreign market also increases, provided the foreign market is sufficiently different from the domestic market to warrant it. Accordingly, affiliates in *large* foreign markets are particularly attractive candidates for R&D activities. This can be seen in those instances where regional markets are standardized, such as in Europe, where R&D laboratories have been established to serve the entire regional market. In other cases the need for markets of sufficient size to support R&D has led to facilities serving informal regional markets. One example is a chemical firm which has divided the world into six geographic areas and has R&D capabilities in each area. The sole purpose of these foreign R&D capabilities is applications development and product modification. According to the executive interviewed, "Their mission is to determine what the state of the art is among the users of a chemical product, duplicate the problems of the users and potential users, solve these problems, and then go into the users' plants and show them what to do." There was also noticeable support among many of the host-market executives

executives interviewed for the proposition that their foreign affiliates are serving distinct markets and, as such, are autonomous business entities deserving their own R&D activities. Accordingly, the primary initiative for the establishment of foreign R&D activities by many of the host-market firms arose from the foreign affiliate itself. In these firms, when the affiliate felt that conditions were "right," it made the decision to establish an R&D group, typically focusing on applied R&D projects.

While all of the companies in the sample interviewed expressed an interest in enhancing their relationship with host-country governments, the host-market firms, with their distinct national-market focus, were particularly sensitive to the importance of such relations. In a number of cases this has resulted in foreign laboratories that are viewed by corporate R&D headquarters as being nothing more than a drag on corporate profitability, with little or no useful scientific or technical output gained or expected from these facilities. More typical, however, is the not quite as voluntary placement of an R&D unit abroad in response to foreign government pressures but with expectations of that unit's yielding results. When established under such pressure, a laboratory can sometimes be integrated into corporate technological activities, often to the extent of continuing to expand the facilities and resources beyond what was originally required. This was particularly true among the pharmaceutical firms, who are often forced to locate clinical testing activities abroad in order to meet host-country regulations. These same firms, however, also cited other, more positive inducements for overseas locations, such as: an ability to get a product started faster abroad without home-country regulatory agencies being involved; the political wisdom of using foreign university researchers, who probably are on local boards of health, for performing clinical trials; and an ability to speed up the drug diffusion process through the use of local researchers who would also, eventually, be potential adopters of the product.

World-Market Companies. The world-market firms appear to be principally concerned with the availability abroad of specific types of skills in particular technical areas. This, of course, is in keeping with their greater propensity to assign new product development responsibility to foreign R&D groups. More than one of the foreign R&D laboratories of the world-market companies was characterized as having achieved a level of competence in a technical area that far surpassed the capabilities of other research groups within the corporation. Accordingly, these laboratories were assigned principal technical responsibility for a specific technical area within the firm. In the words of one R&D executive of a world-market firm: "The key to [our foreign laboratory's] success is the building-up of several good quality areas, above threshold size, to the point where they can be self-sufficient and yet still engage in cooperative research within the corporation. In [one affiliate laboratory] we have about twenty professionals in electrooptics who have the necessary variety of skills to operate as the corporate laboratory in this technical field, with about another twenty to twenty-five people, mostly technicians, working in a supporting capacity." The high quality of this laboratory's work has enabled it to achieve an international reputation and to continue to grow via its attractiveness to scientists and engineers.

The attractiveness of foreign sources of technical skills as an inducement to locate R&D activities abroad was recognized even by a world-market company that has been wedded to the concept of an efficient, highly centralized R&D facility, and that had only an extremely small amount of development work being performed outside of the U.S. According to the executive interviewed, they would consider a foreign R&D location if there was "a local nucleus of good people in a particular technical area, and if this nucleus was large enough to be worthwhile."

A second reason given by the world-market firms for locating R&D abroad is the ability to gain access and contacts with foreign scientific and technical communities. As one American R&D executive put it: "Our Swiss lab is really a European lab, and through it we gain access to all of Europe." For some of these firms, the value of foreign "listening posts" is considered quite high: "Overseas laboratories give us a window on a different world; the opportunity to tap new and different ideas. This has been important in the past but will be more important in the future. As we move away from World War II, recognizing the relatively small proportion of the world's population we [i.e., the United States] have, it's unreasonable to assume that we're going to dominate world R&D as much as we have in the past. Accordingly, we will be increasingly faced with the need to transfer foreign technology back to the United States."

Universities were frequently mentioned as an important means of gaining access to the foreign scientific and technical communities that are of such great interest to the foreign exploratory laboratories of world-market companies. Every one of the world-market firms stressed the need for a strong local university system as a prerequisite for choosing an overseas location for R&D. In the words of one executive, "Environment, the ability to call upon colleagues, the ability to call formal and ad hoc meetings in the immediate physical proximity of the laboratory are all essential ingredients of a desirable location to do R&D. We couldn't for example, do these things in many developing countries because of the lack of access to other institutions in the immediate neighborhood." R&D laboratories of foreign affiliates that establish ties to the local university system often become major entry points into the corporation for the graduates of local technical institutions.

Obstacles to Foreign Locations for R&D

Home-Market Companies. The criteria used in selecting a foreign R&D location reappear among the reasons given by firms for not doing R&D abroad. Among the home-market firms, relatively little is sold in foreign markets; what is, is not significantly different from what the firm sells in its home market. Hence, the home-market firms have regarded their foreign sales as not requiring any further R&D beyond what had been performed for the original market. In most cases these firms were dealing in commodity-type products whose overseas markets were viewed as direct extensions of their domestic businesses. In fact, one company producing "high-volume, nondifferentiated, petrochemical-based, commodity chemicals" had retained its home-market perspective despite having yielded to competitive pressures by manufacturing abroad. The firm had decided that it was less profitable to export from the United States than to produce in the European Common Market because of the need to avoid trade barriers and the existence of lower European wage rates. But since it was essentially a home-market company it did not embrace a multinational corporate philosophy, and made no attempt to develop new technology especially for the foreign market.

Although home-market firms tend to have high exposure in the developing world because of their extractive operations and component manufacture, they have not located many R&D activities in these countries because they typically do not refine raw materials or sell products in these markets. What products are sold in these markets tend to be simple and standardized, not requiring any further R&D.

Another reason widely given by the home-market firms for not locating R&D in the developing countries is their perception of a lack of qualified scientists and engineers to staff a laboratory. Because of an inadequate supply of qualified scientists and engineers, it would be particularly difficult to assemble teams sufficiently large and diverse to achieve a "critical mass" necessary for effective research and development. Since most of the R&D performed by these firms is process oriented, they typically require a greater range of skills and closer association with operating facilities than do companies performing more product-oriented research. This can magnify the need for adequate numbers of trained personnel and reinforce the notion of the effectiveness and desirability of a single centralized R&D effort, most typically located with the parent firm.

Host-Market Companies. Among the most commonly cited reasons for not locating R&D activities abroad given by the host-market firms was a recognition of the increasing cost of doing R&D in foreign locations. This concern over costs was viewed by a number of host-market firms as a sufficient reason not to locate R&D abroad or expand existing facilities there. The perceived adequacy of parent company's R&D in addressing the needs of developing country markets further reduces incentive to initiate R&D activities in these nations. In fact, one host-market firm, which recognizes the need for local R&D efforts to address local product or market needs, sees its future expansion of R&D activities coming largely from a centralized facility that will serve all or most of its overseas R&D activities at far less cost than individual laboratories built for individual local markets.

A third reason arose from a desire to keep exposure low in the face of political instability abroad. This concern included not only a fear of nationalization and the resulting loss of investment but also the difficulty of inducing company researchers to go to a country where personal safety was uncertain. Somewhat ironically, the most frequently mentioned example of such a situation was not a developing country, but Italy. Several of the companies mentioned that they had recently been unable to send R&D personnel to Italy because of the political violence there and, as a result, would certainly not consider Italy as a candidate for future expansion.

World-Market Companies. Perhaps the major inhibiting factor to the establishment of overseas R&D activities by most of the firms interviewed, and world-market firms in particular, is their emphasis on the efficiency of centralized R&D facilities. Simply stated, "Setting up R&D around the world is an expensive way to do R&D." Another firm emphasized that "The problem is to build up a threshold group in something and that's hard enough to do. When you're remote, removed [i.e., without a direct manufacturing link] from the corporation, it becomes critical!" For most of the firms in the world-market grouping, this need for a threshold size for an R&D laboratory—a laboratory of sufficient size to encourage communications, diversity of skills, and sophisticated instrumentation—precluded the developing countries from further consideration as potential locations. The firms interviewed believe that the necessary supporting social and institutional infrastructures are not present in the developing countries; even if a research team of adequate depth and breadth could be assembled in a developing country, some executives suggested that such a concentration of talent would seriously detract from the country's own scientific and technical efforts.

ENTRY MODES AND GROWTH PATHS

Foreign R&D activities of transnational companies arise in three ways: evolution from foreign manufacturing or market-service operations, either wholly or jointly owned; direct placement of R&D activities in a foreign location as an extension of the parent company's R&D program or as a cooperative effort between two or more companies; acquisition of existing foreign R&D activities. Since a firm's market orientation and the mission of a foreign R&D group have been shown previously to be related to location of the R&D activity, it would appear reasonable that they would also influence the method by which a firm establishes R&D abroad and the subsequent development of that R&D group.

Entry Modes and R&D Mission

Thirty of the wholly owned foreign R&D facilities of American transnational firms had missions involving new product research, while forty did not. The most striking difference between the two groups of wholly owned laboratories was the stronger reliance upon direct placement to establish groups whose mission included new product research responsibility (37 percent) as compared to establishing groups whose missions were limited to applied R&D (12.5 percent). Acquisition of existing foreign R&D laboratories accounted for about 25 percent of the R&D activities in both groups (see Table 1.7).

Table 1.7. The Use of Different Entry Modes to Establish Foreign R&D Groups with Various Missions

	Entry Modes		
	Evolution from Technical Services	*Direct Placement*	*Acquisition*
Laboratories whose mission includes new product research	11	11	8
Laboratories whose mission is limited to applied research and development	25	5	10

Source: Interviews.

Note: The data in this table represents the 70 foreign R&D facilities that were wholly owned by American transnational corporations and where both entry mode and mission could be ascertained.

Entry Modes, Growth Paths, and Market Orientations

Evolution. As the discussion above suggests, there is a widespread pattern of foreign R&D originating as technical support activities for manufacturing or marketing. While not generally considered as R&D, these technical support activities are performed by a concentration of technically adept individuals whose interests often expand beyond the narrow boundaries of their organizational mission. To the extent that the technical group can gain formal recognition of its expanding interests and capabilities, its official mission often evolves from support to one of more responsibility. This frequently occurs through the group's taking on projects not officially assigned it until it wins acceptance for technical contributions. As a result, there exists a continuum of R&D missions ranging from quality control and technical service through materials testing, raw material adaptation, process adaptation, product adaptation, and applied R&D, to new product development and exploratory research. One European firm has suggested that the typical evolutionary process in its laboratories starts with five to ten years of quality control and technical service, evolving into modifications of components and materials (at which level the lab may remain indefinitely), evolving, if the market requires, into new product development.

Home-Market Companies. Because of their domestic orientation, it is only in unusual circumstances that it is possible to find a home-market company with a foreign R&D group. This occurred twice in our sample, and in both cases the R&D activity was initiated as a technical support activity attached to a foreign extractive operation. Over time, as the sophistication of the group increased, and as the local scientific and technical infrastructures developed, the mission of these groups evolved to include a limited amount of applied R&D. It was definitely not intended, however, that these foreign facilities would ever become significant R&D activities in their own right within the corporation. This was illustrated in one of the firms by the shifting of projects from the foreign facility to the corporate central laboratory whenever a project exceeded a certain size.

Host-Market Companies. Evidence of evolution in the capabilities and missions of foreign R&D groups was abundant among the host-market companies. A British chemical firm executive put it quite simply: "Technical service is attached to our foreign manufacturing affiliates and, therefore, some developmental work, reflecting our product lines, is always beginning around the manufacturing activities at our foreign plants." The experience of an American chemical-based firm is also instructive in this regard: "Our firm recognized the importance of the European market and built a plant in England, followed by a plant in France. As we began to produce, it become necessary to have technical assistance to manufacturing. Then, technical service groups were established in Germany, Spain, and a few other countries to serve the peculiar characteristics of these markets. Our group in England was trying to tailor subtleties of existing products for the European market. This work continued for some time until, in the late 1960s and early 1970s, it became necessary to conserve corporate resources and cut costs, and so we reduced our activities in England to the point where, today, we do very little R&D overseas." This latter example presents an illustration of the proliferation of foreign technical activities in support of marketing and manufacturing activities, and, as evidenced by the British lab, the evolution and devolution in the foreign laboratories' missions.

A third chemical-based firm with substantial foreign R&D activities experienced an evolution of its overseas R&D activities from a marketing, rather than a manufacturing, orientation: "At our firm, there is a fairly predictable evolutionary pattern associated with foreign R&D activities. When [we] enter a foreign area we typically start with a broker and establish a sales office. Technical problems associated with serving the market create the need for hiring a technical service person and this, in a sense, marks the birth of the R&D group. As sales volumes sufficient for local manufacturing are reached, a plant is established and manufacturing initiated. Technical people are then needed to help new products, and applications groups are created. This evolutionary process typically takes five to eight years, but, in the case of our largest and most sophisticated overseas R&D group, it took ten years."

An energy-based firm's Canadian laboratory has evolved from providing technical assistance to marketing and manufacturing to adopting U.S.-originated technology for the colder Canadian market to developing its own products for cold-weather markets. This evolution has taken five to ten years.

A pharmaceutical firm discussed laboratory evolution in terms of a Latin American laboratory it had originated because of a need to be close to the raw material and production sources for a major pharmaceutical product. "Over a period of fifteen years, the lab has evolved to the point where it is now performing development work of a limited nature. Aside from manufacturing ties, this [Latin American laboratory's] evolution was facilitated by unique Latin American markets and the fact that product registration requirements there are low. Some product development work for the Japanese market is performed in Latin America, for example, because of these low requirements."

The international differences existing in government product regulation led to an evolution in overseas R&D activities in another American pharmaceutical firm: "[One of our European laboratories] was developed to do the local toxicology work required by [host] government regulations. Since then, in an effort to reduce research duplication, this laboratory has been given much of our worldwide toxicology work." While this facility did not evolve directly out of manufacturing operations, the firm did take advantage of already existing operations in establishing the facility. This group's mission evolved over two years from specialized work in toxicology to include some fundamental research work.

A second European laboratory of this same firm was preceded by a local subsidiary that has sold chemicals, agricultural products, and pharmaceuticals abroad for many years. This firm ships active ingredients manufactured in the United States, overseas for formulation. To accommodate this work, a formulations laboratory was established. Over a period of years, however, this facility's experience with European formulation practices, as well as its ability to service European markets, has made it a valuable R&D center in its own right, and its mission has expanded. According to the executive interviewed: "Occasionally a major change in capabilities is necessitated as a result of different supply situations. This calls for modification of an existing product. You could, of course, send the materials back to the United States for modification, but then you would wind up using different tableting equipment or different capsule-filling equipment, and it just really doesn't make sense to do additional translation of things of that type." An important consideration in establishing this R&D activity was the pharmaceutical firm's ability to take advantage of existing foreign operations to reduce the burden of the fixed costs of a new facility. As one executive put it: "An important problem for any satellite lab is that a certain amount of services, such as heat, light, maintenance, security, and the like, are needed. If the lab is too small, the fixed costs per person of these services will kill you!" Hence, there is strong impetus to co-locate R&D activities at existing facilities.

The continued growth of foreign technological capabilities has also provided an impetus for laboratory evolution. An American firm producing machine tools, having had manufacturing operations in Europe for more than forty-five years, noted that until ten years ago the United States was technically far enough ahead of Europe for this firm to view Europe as a market for extending the life cycle of products originally introduced in the United States that were in the waning portion of their domestic life cycle. At that point, the firm's R&D for foreign markets was basically modification to meet peculiar characteristics of the European market. At present, however, the technological gap between the United States and Europe is minimal. As a result of this, and while the dollar was pegged, this firm started to move R&D into Europe. That strategy, however, has been reversed since the dollar was floated, making R&D conducted abroad more expensive. Accordingly, this firm has recently reduced its foreign research in all labs except one and has closed laboratories in Germany and Belgium when their companion manufacturing facilities were closed.

Evolution of foreign laboratory capabilities and mission was also illustrated by a European chemical firm that established an R&D laboratory in the United States to support manufacturing and marketing operations, with any new products and processes remaining the responsibility of the European parent. The technical capacity of the American laboratory increased as the company grew and as an engineering development group, based upon European know-how, was added. As the composition of U.S. products and the U.S. product line began to differ from the European, different scientific and technical capabilities were required by the U.S. lab. Consequently, the U.S. laboratory is now self-supporting in product development; and, while new processes still tend to originate in Europe, this lab is now capable of some process development as well.

A further example of laboratory evolution was provided by an American pharmaceutical firm which, although it had no manufacturing facilities in France, had to locate an R&D group there to perform the clinical work necessary to sell its products in France. The high quality of the local scientific and technical personnel, however, and the importance of the Common Market will, according to an executive in the firm, undoubtedly allow the French laboratory to evolve into performing product discovery research.

The importance of evolution as a growth path in determining laboratory missions and capabilities in host-market firms is indisputable, given that it is the process followed in more than one-half of the growth paths of the foreign R&D laboratories of host-market companies. This should not be surprising, given the inclinations of the host-market companies to get as close to their customers as possible in order to tailor their products to the needs and resources of the markets they serve.

World-Market Companies. World-market companies typically do not share the evolutionary pattern, from technical services to higher levels of responsibility, that was found among host-market companies. Rather, they tend to create, by direct placement, their foreign R&D groups. However, even within groups created through direct placement there is some evidence of evolution. As an example, an R&D executive in a pharmaceutical firm explained how his firm would approach the direct placement of a foreign laboratory to do basic research: "The lab would be limited to working in one technical area only, for the first four to five years. At that point it might be possible to extend this effort into another technical area. This incremental approach is more manageable and allows the "critical mass" to be developed easier and quicker." Thus, although this firm relies on direct placement of R&D resources for the creation of foreign R&D groups, once in place these groups experience an evolution in their mission and capabilities.

A further example of evolution in the foreign R&D activities of world-market companies can be found in three overseas laboratories, which began as international "listening posts," and now perform new product R&D for world-market companies. In one electronics-based firm, this was done in the late 50s and early 60s in Europe and in 1960 in Japan. These labs have both since switched to an emphasis on exploratory and new product R&D. In a second electronics-based firm, "the decision to place a laboratory in Europe was partly to establish a listening post overseas in the pre-jet travel era and partly to do research. At present, this lab is about 99 percent research." This R&D manager claims that, because of the ease of travel and communications, the idea of a listening post in today's world is considered outdated. In each of these cases, however, an R&D activity that was originally directly placed in a foreign location to perform a relatively limited task has evolved into considerably more ambitious undertakings.

Direct Placement. The term "direct placement" is used in this study to represent the decision to create a foreign R&D group to perform specific scientific and technical missions other than simply supporting local manufacturing and marketing operations. In the case of direct placement the foreign R&D effort is intended, from its inception, to do R&D. While the type of R&D such labs perform can vary considerably, the important point is that they are specifically established to do R&D, and R&D only, at a specified (fairly high) level, at inception. Although evolution is a more common growth path for foreign R&D activities, direct placement is by no means unusual. In fact, one American R&D executive emphatically stated that, in general, "Most overseas R&D units do not evolve but, rather, are created to satisfy particular needs. This is different from the domestic American experience where laboratories do tend to evolve. When you are considering overseas R&D operations, however, you are starting with a well-established American company. Its overseas operations presents new missions which it must perform. Accordingly, an R&D unit placed overseas is perceived as an answer to the new need or specific function and are usually established full-blown."

Home-Market Companies. Those firms whose primary orientation is to serve their home-country market typically have no incentive to create, by direct placement, a foreign R&D activity. As was noted previously, these firms tend to "suffer" foreign R&D groups only to the extent that they are necessary for support of their foreign extractive operations. To the limited extent that these firms sell any products outside the home market, they tend to be direct extensions of the products offered domestically and thus require no additional

R&D. Accordingly, it came as no surprise to find no instances of direct placement of foreign R&D by the home-market companies in our sample.

Host-Market Companies. The earlier discussion of motivation for the establishment of foreign R&D activities by host-market companies identified the need to be close to the local markets, the relative autonomy of their foreign subsidiaries, and their relationships with host-country governments as being particularly important. While these three motivations are certainly not mutually exclusive, they play roles of differing significance in explaining the origin of those foreign R&D activities of host-market firms that were established through direct placement.

Perhaps the most compelling need for foreign R&D activities by a host-market company occurs when the firm is attempting to serve a unique market that is also geographically distant. Direct placement in such cases is extremely useful because it offers the ability to quickly place R&D resources close to the target market. A European company attempting to serve the complicated U.S. construction market, for example, entered the market by acquiring a well-known U.S. firm and then, since the acquired firm lacked its own scientific and technical resources, created a U.S. laboratory through direct placement. In this way, an R&D group of 500 people was established quickly.

There are, of course, potential problems associated with the direct placement of foreign laboratories by host-market companies when the "fit" between the laboratory and the foreign manufacturing/ marketing activity is not perfect. This recalls the admonition, offered by one executive, that in order to be successful a foreign R&D laboratory "needs to have something [i.e., some operating entity] to transfer [its] R&D results to." Typically, when laboratories evolve, this "fit" occurs naturally. This is not necessarily the case with direct placement. As an example, a European petroleum-based firm established a laboratory in Germany through direct placement with the express purpose of attracting that country's best scientists. It succeeded, but the laboratory failed anyway because it was disconnected from the firm's local operating affiliates. Its mission was to respond to long-term corporate needs, but in the short run, its capabilities could not be justified in the face of a tight budget. Consequently, the laboratory was closed down for lack of recognized contribution to the company.

A second major motivation for host-market firms to directly place R&D activities abroad is the relatively high degree of autonomy that their foreign subsidiaries enjoy. In many of these firms, the local management of the foreign subsidiary is responsible for making the decision to initiate an R&D program. Although such decisions are often subject to review and reversal at higher managerial levels within the international corporation, the overruling of the management of host-market subsidiaries is uncommon. One food industry executive explained the degree of autonomy this way: "The decision to have an R&D unit, or not, is up to the country manager and is pretty much a function of market size. If a decision is made, a development group is created. At present, for example, if the Germans wanted an R&D unit they could have one—it's their decision." In this firm, once the go-ahead is given, an R&D group is generally expressly created rather than allowed to grow through evolutionary patterns. A chemical-based firm reported that "our foreign R&D activities are entirely funded by the foreign subsidiaries and the decision to have a lab or not, program development, project selection, and monitoring, are all the responsibility of the foreign subsidiary." Given this high degree of autonomy, a number of the foreign affiliates of host-market companies have created, relatively quickly, R&D groups through direct placement in order to gain the services and prestige associated with having their own R&D group.

Government pressure has, in a number of cases, resulted in direct placement of host-market company R&D laboratories overseas, in an effort by the firm to appease the host-country government. One example involved a European company that created an R&D unit in Spain through direct placement in response to the Spanish government's refusal to allow repatriation of royalties from technology licensing. The decision to invest funds in this venture was made despite the firm's lack of confidence in the outcome, yet the capabilities of the laboratory have evolved to the point where there is very little difference between the research performed there and at headquarters.

World-Market Companies. Direct placement appears to be an even more important means of establishing foreign R&D activities for world-market companies than it is for host-market companies. As discussed earlier, world-market firms establish foreign R&D laboratories, with a substantial commitment to new product research, in locations where they have access to scientific and technical communities and pools of talent in particular technical areas. It is thus not surprising to find that a substantial number of the foreign laboratories of world-market companies stand on their own, distinct from any operating activities the firm may also have in a particular country. Examples of this are the three foreign laboratories discussed earlier, now performing new product R&D for world-market companies, which originally began as international "listening posts" and were directly placed in sites appropriate for that mission. A further example is provided by a European firm that has a number of foreign laboratories, including four with a heavy research emphasis. According to its executives, whereas the size and nature of a development laboratory is restricted by the local market and its demand, the activities of each research laboratory are determined by the nature and size of the scientific community from which it can be drawn. This firm's development labs tend to follow a predictable evolutionary pattern from quality control and technical service, through minor modifications, to ultimately some product design responsibility. Development activities are *always* attached to a specific factory so that they can mesh with the needs of the operating unit. Its research laboratories, on the other hand, are not located near the firm's plants, and result from corporate rather than divisional initiative. Each research laboratory has a "complete task" fitted into a total research program for the firm as a whole. The four foreign research laboratories are equal in status and are essentially self-contained, even in methodology.

Acquisition

Home-Market Companies. Because of their home-market orientation, none of the firms in this grouping had any experience with, or expressed any inclination to consider, the acquisition of foreign R&D resources.

Host-Market Companies. The only firms that had employed acquisition as a means of establishing R&D activities abroad were in the host-market grouping. According to one R&D executive, "Any time we want to get into something new, we buy someone; it's quicker and safer!" In slightly less than 30 percent of the American host-market observations gained during our interviews, overseas labs were established through the acquisition of a foreign firm. As a general rule, the acquisitions were made for financial and/or marketing considerations, and the R&D group was not directly involved in the acquisition process. As one American R&D executive put it, "Had we not acquired European firms with laboratories, we wouldn't be in Europe." Having once acquired a foreign lab, however, the parent company generally allows it to enjoy its autonomy and continue its work if its products are significantly different from those of the parent firm, or to selectively expand its resources if its projects can complement the existing R&D of the parent firm.

Examples of corporate acquisitions of foreign R&D laboratories with different product offerings include a pharmaceutical firm that bought an ophthalmological laboratory, a chemical company that owns a pharmaceutical laboratory, and an industrial commodity producer that owns a European cosmetics company. While in all of these cases the acquired firm has been allowed to continue its own R&D, the new parent company has involved itself in a variety of ways, ranging from establishment of quality control to actually expanding the R&D resources and assigning worldwide ("lead division") R&D responsibility to the foreign affiliate.

In those instances where the acquired firm is in the same business as the parent, attempts are usually made to reduce R&D duplication. In one pharmaceutical firm the acquired foreign laboratory was asked what it thought it could do best, and then its research program was melded into the total corporate research program to complement, and be complemented by, the original corporate R&D activities. An energy-based company began "sharing" research results with a foreign energy-based company prior to its acquiring the foreign company. The purpose of the acquisition, in this case, included gaining technical expertise, a desire to enter the foreign market on a large scale, and financial considerations. In the post-merger period, after some initial continuation of the relatively autonomous relationship that characterized the premerger relationship between laboratories, the R&D headquarters established joint research projects to pool expertise.

A firm in the automotive components industry has experienced an unusual situation in the postacquisition period as a result of the world energy shortage. This firm acquired two automotive R&D laboratories in Europe, based on nontechnical financial interests. For some time these two acquired laboratories were left alone by the American R&D headquarters, which had little interest in the "small-car" products being worked on in these laboratories. As a direct result of the energy shortage, however, the executive interviewed foresaw "a major change coming. We see major changes in automobile components and car size in the United States in the not-to-distant future. Therefore, we forsee much greater interchange of R&D between us and our foreign labs with much of their experience with smaller cars being transferred back to us." This R&D director estimated that it took six to seven years for the acquisition to really take effect, as there had been some resistance on the part of the acquired laboratories to giving up their independence. Now that market demands are becoming similar, however, substantial impetus exists to technically consummate the merger.

Two major problems encountered in the acquisition of foreign R&D laboratories are the failure of corporate R&D to be involved at anything more than a cursory level in preacquisition planning, and difficulties with control relationships in the postmerger period. Among the firms interviewed for this study, there was no instance where R&D had been seriously included in the decisionmaking process leading to the acquisition. While the R&D managers had, in some cases, been asked to comment on the technical soundness of an acquisition candidate, they usually bemoaned their exclusion from the acquisition planning process.

The postacquisition control problems reported by firms in the sample typically involved difficulties in introducing an appropriate amount of control over the acquired foreign R&D laboratory so as to coordinate it into the new parent's R&D program. As an example, one French firm with an American affiliate noted that its American affiliate would not accede to its demands for tighter coordination. Furthermore, the U.S. laboratory officials cannot speak French, causing communications problems as well. The French executive interviewed presented the slightly ludicrous picture of a meeting where eighteen Frenchmen were trying to speak English in order to satisfy the two Americans present. In another example, an American pharmaceutical firm spent several years trying to wrest control of a European firm from the entrepreneurial founder-owner from whom they had bought it. This individual stayed on in a "consultancy" position and tended to run the business in much the same way he had prior to selling out; control was only really established after the founder-owner retired.

World-Market Companies. There were no instances of world-market firms in the sample acquiring existing foreign R&D laboratories. Considering these firms' use of foreign R&D activities as a means of attracting foreign scientific and technical talent in particular research areas, it is not surprising that they have not employed acquisition as a method of establishing R&D activities abroad.

SUMMARY

The information gathered and presented in this chapter shows that transnational corporations actively pursue R&D abroad.

While the sample of firms interviewed was largely composed of companies with foreign R&D experience, the discovery of an average of almost four foreign R&D groups per firm is considerably more than the literature and "conventional wisdom" indicate. Somewhat surprisingly, European firms were found to have more R&D activities abroad than do U.S. firms. While this may reflect geographic and cultural proximity as much or more than philosophical differences, it was another finding that was not anticipated. Of particular note was the high level of European corporate R&D performed in the United States and the willingness of European transnationals to pursue new-product R&D abroad.

Most of the foreign R&D of the transnational corporations is to be found, as expected, in industrialized countries, and is primarily development-oriented. However, over thirty different countries were cited as hosting the foreign R&D activities of the firms interviewed, and a substantial number of foreign laboratories were pursuing new-product research.

The findings lead us to conclude that there are more similarities than differences among the American and European transnationals. This results from similar market-orientations, which are more significant for R&D than national origin. These market-orientations were found to be quite useful in differentiating the missions and choice of locations for the firms' foreign R&D activities. Host-market firms in our sample are by far the most active pursuers of R&D abroad, while the laboratories of the world-market firms were most likely to have missions involving new-product and exploratory research.

Market orientations also helped explain the factors that various firms considered to be important in making the foreign R&D location decision. To the extent that they had foreign R&D activities, home-market companies chose sites close to their foreign operations in order to technically support these operations. Host-market companies, on the other hand, are considerably more concerned with market-related matters, such as proximity to local markets and the need to respond quickly and tailor R&D efforts to local needs. World-market firms are primarily interested in gaining access to specialized groups of foreign scientific and technical talent. Table 1.8 provides a summary of those factors considered in the foreign R&D location decision.

Slightly more than one-half of the foreign R&D groups of the American transnationals were established through the evolution of skills, capabilities, and missions, from an initial emphasis on technical support to manufacturing and marketing to applied R&D.

Table 1.8. Important Criteria for Considering or Not Considering Overseas R&D Locations

	Home-Market Firms	Host-Market Firms	World-Market Firms
Important criteria for considering an overseas R&D location	1. Proximity to operations	1. Proximity to markets	1. Availability of pockets of skills in particular technical areas
	2. Availability of universities	2. Concept of overseas operations as full-scale business entities	2. Ability to access foreign scientific and technical communities
			3. Availability of adequate infrastructure and universities
Important criteria for not considering overseas R&D locations	1. Lack of sophistication of products sold in the developing countries	1. Increasing costs of doing R&D overseas	1. Economies of centralized R&D
	2. Lack of qualified scientists and engineers	2. Economies of centralized R&D	2. Difficulties in assembling R&D teams
	3. Economies of centralized R&D		

Foreign laboratories with new product responsibilities, however, are much more likely to be established through direct placement rather than evolution.

The market-orientation of the parent firm was also found to be quite useful in understanding a firm's choice of methods of establishing R&D abroad. This was evidenced by the heavy reliance of home-market companies on evolution in the establishment of foreign laboratories, while host-market firms engaged in a variety of entry modes, and world-market firms tended to favor direct placements.

NOTES

1. The appropriateness and utility of activity categorizations in R&D has been a topic of continuing discussion in the literature on R&D. The categorizations employed in this study are not exempt from the shortcomings inherent in any categorization scheme. Rather than aspiring to great precision, we have attempted to use our categories to capture what is happening in these foreign facilities. A lab described as having fundamental product research within its mission merely indicates that the R&D group is expected to go beyond product and process modification or adaptation and to actually develop new products.
2. Although the data gathered in the European interviews do not allow a detailed breakout of activities, by location, such as that presented in Table 1.4, the conclusions reported here are supported by the descriptions obtained in the European interviews.

Chapter 2

The Management of R&D Activities Abroad

Organizational structure and control systems are important determinants of almost all aspects of R&D behavior, including decisions on location, selection of an entry mode, the choice of a particular growth path, and coordination and communication procedures. This chapter discusses the manner in which foreign R&D activities are controlled and coordinated by the parent organization and the relationship between coordination and control practices and R&D resource allocation.

MANAGERIAL STYLES

There are a number of ways by which foreign R&D activities can be coordinated and controlled. It is likely that each organization exercises its coordination and control responsibilities in a fashion virtually unique to itself. (Appendix A presents some idea of the variety of organizational structures for R&D management found within the firms interviewed. There appears to be no one generally accepted pattern for organizing foreign R&D activities.) Even so, some structure can be given to these various styles, and from this structure a taxonomy can be derived that shows the relationships between the parent and subsidiary and that elucidates the ways in which coordination and control mechanisms affect R&D activities abroad.

Accordingly, it is necessary to develop a taxonomy that accurately reflects managerial styles. Just such a scheme was developed recently by DeBodinat;[1] by using it, each of the firms interviewed can be assigned to one of five categories:

> *Absolute Centralization.* Commitment is imposed by the parent on the subsidiary.
> *Participative Centralization.* Commitment is reached as a result of negotiation between parent and subsidiary. Parent decides, subsidiary gives advice or proposes decision.
> *Cooperation.* Commitment reached by agreement between parent and subsidiary approaching equals.
> *Supervised Freedom.* Commitment established by subsidiary's decision. Parent may express opinion or make suggestions.
> *Total Freedom.* Commitment established by subsidiary and automatically accepted by the parent.

Figure 2.1 illustrates the nature of the relationship between parent and subsidiary for each of these influence relationship categories.

The nature of influence over R&D within a firm was determined, for this study, by examining the interview records and attempting to gauge from these the relative roles of corporate R&D headquarters and noncorporate R&D management in matters of resource allocation, program initiation, selection, monitoring, and termination of projects, and communication and coordination activities.

More than one-half of the American-based transnational companies interviewed were categorized as having some degree of centralized control over their noncorporate R&D activities, while the remainder displayed some form of decentralization. Among the European firms, slightly more than one-half exhibited decentralized influence patterns (Table 2.1).

In no case was it felt that there was equality between parent and subsidiary and, therefore, no firms were assigned to the category entitled "cooperation." This should not be taken to suggest that no firms were attempting to act in a cooperative fashion, but rather that in every case there was sufficient inequality in the parent-subsidiary R&D relationship to place the firm on one side or the other of the freedom–centralization continuum.

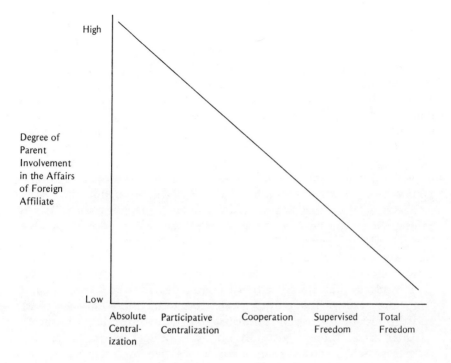

Figure 2.1. Relationship between degree of parent involvement in the affairs of the Foreign affiliate and the managerial style employed.

Table 2.1. Employment of Various Management Styles in Influencing Foreign R&D Activities by American and European Transnational Corporations

	American Firms	*European Firms*
Absolute centralization	7	0
Participative centralization	11	6
Cooperation	0	0
Supervised freedom	13	10
Total freedom	3	0
Total	34	16

ABSOLUTE CENTRALIZATION

Only seven of the firms interviewed could be characterized as "absolutely centralized" in managing their R&D activities. All of these firms are American, and only three have R&D activities located abroad. Absolute centralization does not appear to be a management style that is easily compatible with the performance of R&D in foreign locations. Of the three firms practicing absolute centralization in managing their foreign R&D activities, two have extremely small development groups in Canada, one of which is due to be terminated, and the third has a minority interest in a foreign joint venture that includes a technical service component.

The firms classified as practicing absolute centralization generally subscribed to the "efficient central R&D facility" school of thought. They tend to separate R&D from the line divisions and to exercise substantial control over the allocation of R&D resources and the flow of information from R&D activities to other parts of the firm. The result is that R&D in these companies appears to be isolated from the rest of the corporation. Communication between R&D and the rest of the organization usually occurs informally. In several of these firms there is virtually no travel budget for R&D (particularly research), and the principal interactive mechanism between R&D and its corporate constituents is the personal involvement of the R&D manager. None of the three companies in this category which have foreign R&D activities can be considered as having a strong science-orientation.[2]

Home-Market Companies

Three of the seven home-market companies interviewed exhibited absolute centralization in their management of noncorporate-level R&D activities, as demonstrated by their lack of foreign R&D activities. Two of these firms are in the nonferrous metals industry, and one produces bulk, or commodity, industrial chemicals.

One of the three firms has a minority interest in a foreign company that has its own R&D group, but, while there is frequent informal contact between the R&D director of the joint venture and the corporate R&D director, and while R&D staffers from both the joint venture and headquarters meet once every three years to share information, the corporation treats the joint venture as merely another customer on matters of proprietary research. This same firm supports some process research performed by a foreign industry research cooperative because of the unique aspects of that research, but the support does not constitute a significant portion of its research budget and does not affect its own research program.

In general, the firms in this group tend to have a single central R&D facility and regard R&D for overseas markets as being unnecessary because any products they are selling overseas are direct extensions of domestic production runs. One of the three firms actually had two domestic laboratories until four years ago, when it decided to merge the two and separate out technical services (which was transferred to sales) and quality control (which was transferred to manufacturing) in order to achieve an isolation of R&D from day-to-day operations. This company's research program is developed solely by the research division itself. It solicits other opinions and considerations in determining its R&D program by saying, "This is what we are going to do, for the following reasons. Don't you agree?" The ultimate decision on what to do, however, rests within the R&D group. There is no high-level corporate committee giving instructions and/or guidance to the research division. In addition, according to the executive interviewed, "at regular intervals, the product managers within the firm are invited by R&D to discuss related work. In all cases, however, while the Research Division welcomes suggestions, it reserves the right to make its own decisions."

To the extent that the firms in this category have any foreign manufacturing activities (and this does occur to a very limited extent when the exporting of domestically produced products becomes so costly that the firm chooses to duplicate its products abroad in foreign manufacturing operations), there does not appear to be much interaction between the local technical service groups attached to the foreign manufacturing sites and corporate technical headquarters. One of the firms in this category had a technical service laboratory in Switzerland that served European export customers before the firm began producing there. When European production began, the technical service group in Switzerland was disbanded and relocated at the plant sites. Presently, other than for occasional analytical work, there is little interaction between these technical service groups and corporate R&D, although the technical service people do interact with their domestically based technical service counterparts.

The informality of communications between R&D activities in companies characterized by an absolute centralization management style concerned one of the R&D executives in the home-market oriented company mentioned earlier that was supporting foreign R&D in both a joint venture it participated in as well as an industrial research cooperative it belonged to. His firm very definitely relied upon informal communications at the top of the R&D hierarchy, and he indicated that while the firm's president felt such informal relationships were both workable and adequate, he, himself, felt they were workable but not necessarily adequate. He mused that in many instances, despite the critics' claims, multinational companies couldn't even control themselves, much less anyone else.

Host-Market Companies

Three host-market companies were among those characterized as having an absolute centralization R&D management style, but only one had a foreign R&D laboratory. One of these firms was in the packaging goods industry, one was a brand name consumer products company, and the third was an international engineering contractor.

The firms in this grouping all had a centralized R&D facility within the United States but, reflecting their host-market orientation, appeared to be considerably more sensitive to the needs of foreign markets than did the home-market companies described previously. The strength of the central research laboratory in these firms was illustrated in the project selection and monitoring practice found in one of them: "Any time a foreign subsidiary wishes R&D work performed by the central laboratory, they must submit a New Product Development Request which initially requires local approval, then goes through the regional headquarters, after which it is forwarded to corporate headquarters in the United States. At corporate headquarters, the request goes to R&D for determination of viability and then is sent to headquarters of International Operations for approval. The budget for international R&D is allocated back from R&D to International Operations and, within International Operations, it is allocated to subsidiaries on the basis of projects. Thus, resource availability is clearly a top-down decision from corporate headquarters." The focal point of R&D coordination in this firm is within the International Operations group, which has a development director responsible for each of five worldwide geographic areas. These development directors act as the channels of communication between R&D, Marketing, and the field. All dissemination of information from R&D to the foreign affiliates must go through development directors who selectively filter the information prior to transmitting it. Discretionary responsibility is delegated for even minor product changes within this firm; all such changes must be approved by corporate headquarters. This is partly a reaction to headquarter's perceptions that "unapproved tampering" with the product in an effort to reduce costs has led to significant sales declines in overseas markets in the past. Coordination of technical activities and dissemination of information are largely accomplished through personal communication which, according to one executive, "is somewhat ineffective because of the lack of technical sophistication in the overseas operations and, as a consequence, R&D's perception of having nothing to say to them."

In a second firm in this group, the strength of the central laboratory was evidenced by a recently adopted policy charging the central corporate laboratory with responsibility for all new products and processes, and relegating the divisional engineering and development personnel to maintaining the present situation. The budget for this company's central R&D activity is determined by a fixed percentage of corporate sales and thus is not project-based. As a result, there is some feeling among the individuals interviewed that the central R&D group lacks direction as to what is really needed from it. Also, most new ideas originate from within the research group. Interaction patterns (or the lack thereof) between the R&D group and other parts of the corporation reflect the relative isolation of R&D. There is no formal means of interfacing in this firm, and no formal reports are issued. The R&D personnel travel very little and have very little contact with foreign scientific and technical communities. What interaction with outside groups does occur is largely the result of personal contacts by the group director, who is believed by his subordinates to Le heavily involved in interface activities and travels 10-20 percent of the time. He also acts as an information filter both into and out of the group.

World-Market Companies

Only one world-market company in our sample was found to exhibit absolute centralism in the management of its R&D activities. This particular firm is in the industrial electronics industry. Although it is organized on a regional basis for manufacturing and sales, it has a single R&D group located at corporate headquarters, and a small development group in Canada. The strength of the central laboratory in this firm is revealed in a recent decision to move the Canadian work back to corporate headquarters *because* it is going particularly well! This firm believes strongly in the benefits of a single centralized R&D unit. In keeping with the firm's world-market orientation, all product planning is done for a worldwide marketplace: products are standardized and processes are similar, depending upon market size rather than the local factor endowments of a host country. One reason for the employment of absolute centralization as R&D management style in this firm may be imputed from the comments of one executive who noted that "we need a big enough nucleus in a laboratory to gain technical synergy and new development, and in this industry to be really successful you have to start over and develop new products, not modify existing ones."

PARTICIPATIVE CENTRALIZATION

Seventeen of the firms interviewed were characterized by "participative centralization" in their R&D management. Of this seventeen, eleven were American firms, and six were European. All had experience with foreign R&D activities, and about 30 percent of the U.S. firms' foreign R&D groups had missions that included new product research responsibilities. The participative centralization firms tend to exert strong centralized authority over the funding, programs, and often even project selection decisions faced by the overseas subsidiaries. They also appear, in general, to have a large number of overseas R&D activities and to employ a structured and often sophisticated coordination and control system. There is, however, some evidence of genuine participation by the overseas activities in the management of R&D activities.

There appears to be an equal mix between the choice by these firms of evolution from technical services and direct placement as entry modes for foreign R&D activities; slightly less use is made of acquisition. Those firms in our sample within this category appear to have a strong science-orientation—namely firms within the electronics, pharmaceuticals, and chemicals industries.

Home-Market Companies

None of the home-market oriented companies in our sample exhibited participative centralization as an R&D management style.

Host-Market Companies

Fifteen of the seventeen firms judged to be employing participative centralization as an R&D management style had a host-market orientation; eight of these were pharmaceutical firms, and another six were in chemicals. The relationship between a strong central

R&D coordinating mechanism and foreign laboratories with new product responsibility can be clearly seen among these firms. In one European company, each functional division submits inputs to corporate R&D for possible inclusion into the corporate R&D budget. The R&D group, however, determines its own program, and although this does not have to be specifically related to the operating divisions, its funding comes from charges levied upon the operating divisions. Centralization of research is sought by this firm in order to increase efficiency in serving the needs of its affiliates. The executives interviewed believe that any redistribution of R&D resources toward the divisions would probably increase inefficiency. There is, however, a substantial amount of R&D performed within this firm at the operating division level. Participative centralization in this firm operates as follows: yearly, each operating division laboratory submits its entire R&D program to corporate headquarters for assessment for funding and support; headquarters then provides funds for those portions that are relevant and useful to operating divisions in the firm; division laboratories will modify their programs to get "group sponsorship" and to reduce their own costs; all laboratories report on a monthly basis in the firm's "Research News," with all projects being reported on separately, in the process becoming available to all other laboratories. The guiding principle here is "all have access to all."

A second example of the formality and close coordination of participative centralization was shown by a European pharmaceutical firm where an eight-person group exists within the corporate staff to coordinate all units within the R&D budget, excluding only the laboratories charged directly to production (i.e., quality control and technical service). A central committee consisting of the Directors of Central Research, Central Applied Technology, and the research unit in the Engineering Division, the research directors of the operating divisions and the divisional technology application directors, plus some corporate staff, meets four times a year to review budgets, coordination problems, documentation, toxicology, and publication policy, and to conduct program reviews. Program coordination within this firm is the responsibility of this corporate group. Each year it checks every project above $50,000, representing 80 percent of the total R&D budget, to see if goals are being pursued, commercial feasibility exists for the anticipated results of the project, the patent situation is favorable, the capabilities of the researchers are adequately and appropriately being used or if there is a need for additional expert personnel, and to assess potential future sales.

An American chemical firm employs a more formal participative centralization style:

> For each major product there is a Global Product Coordinator who reports to the Director of Product Research at Corporate Headquarters. These coordinators visit each area every year or two to make sure that the technology is being transferred and to reduce duplication. In determining the corporate R&D budget, a business team is formed for every product and includes representatives of Research, Development, Marketing, and Manufacturing. They decide what R&D is needed during the next year and they approve a budget. This proposed budget is reviewed by the product departments (profit centers) and then goes to R&D Administration. R&D Administration always has to go back and cut this budget. Determination of projects is chiefly made at the business team level and by geographic area laboratories, within the overall constraints of corporate plans for funding R&D. Project monitoring is done at the business team level with worldwide periodic reviews or specific technical research areas undertaken by R&D Administration.

These reviews represent one of the few instances where R&D Administration actually gets down to the project level, although the decisions made at corporate headquarters appear to permeate the entire hierarchy of R&D decisionmaking within the firm.

The high degree of formality in resource allocation and program monitoring that has been shown in several firms exhibiting a participative centralization management style was also seen in an American transportation components producer. This firm's overseas affiliates originally relied upon the U.S. activities for R&D support. Over time, however, R&D capabilities have been spread worldwide. At present, the determination of where R&D will be done is the responsibility of what is called a "worldwide product manager," who reports to a Group Vice President. Within each group there are autonomous product lines for which a worldwide product manager has global responsibility. On his staff is an engineering manager who is responsible for the total R&D portfolio, worldwide, for that particular product. Coordination of R&D is also effected by formally structured semiannual meetings of the firm's global technical experts dedicated to particular products. In addition, coordination is achieved through a Product Planning Identification System in which, each year, each engineering manager must identify, by program, his R&D and engineering development activities. This identification includes a statement of the program's business objective, funding, scope, estimated year of introduction, and estimated year of maturity. Each program must be signed off by the Division Manager and submitted for review and approval to the Manager of

Product Program Development at the Engineering and Research Center, the Group Vice President, the Vice President of Corporate R&D, the Director of Planning, and the Director of Worldwide Operations for the pertinent group.

In some cases, the centralization of R&D authority does not necessarily mean centralization at "headquarters" but, rather, specialization (and thus centralization) at specific laboratories within the firm. As an example, a machinery firm which is attempting to become a world-market firm centralizes R&D on a product at the laboratory where the product was initiated. The originating group works on the basic machine or process with a view towards the world market; modifications to suit local peculiarities are performed later by locally-situated R&D resources. Once a major development gets going, mini-conferences are held on a working level with participants pulled from the firm's worldwide R&D resources. If something is particularly important to a particular market (e.g., the United States), R&D people from that market will be assigned to the foreign R&D group performing the research in order to influence the direction of the innovation. The centralization of influence in decisionmaking is thus strongly felt within any particular product line, but that centralization is also dispersed throughout the firm as a whole.

A second example involves a pharmaceutical firm that reported that it is in the process of reducing the number of clinical trials it performs by centralizing almost all clinical work, excepting the local phase-three trials, in order to obtain a smaller number of well-controlled trials that will be universally accepted. Such centralization does not, however, necessarily mean less research abroad. Some of its earliest anticancer drug trials, for example, are being conducted in Europe.

Participative centralization can thus mean considerable freedom for the subsidiary within overall constraints determined by a centralized headquarters. The chemical firm whose "Global Product Coordinating System" was described earlier as an illustration of a highly structured, formal, centralized control system, also solicits R&D funding needs from each of its overseas areas and incorporates these estimates into the corporate R&D budget proposal which is submitted to the Board of Directors. Similarly, the firm in the transportation components industry, which also employs a highly centralized control system through a worldwide product manager concept, arrives at its proposed corporate R&D budget by totaling up the R&D budgets for its product lines. Hence, the real centralization in this particular firm is exerted a step or two down in the corporate hierarchy below the R&D staff at corporate headquarters, but nonetheless represents more emphasis on centralized than on decentralized decisionmaking.

Similarly, the centralized nature of participative centralization was seen in a pharmaceutical firm where the ranking R&D executive noted that "management's responsibility is to decide what areas the firm's R&D efforts should be applied in, and to bring the right people together in teams." On the other hand, he demonstrated the participative nature of "participative centralization" when he went on to say that "What these teams actually do, however, is their responsibility. If management is going to play a strong hand at this level there is no role for the senior scientist. Accordingly, the actual project selection decisions at (our foreign laboratory) are made by the senior scientists at the bench."

Within firms characterized by a participatively centralized management style, communications between R&D and other corporate functions appear to exhibit a high degree of formality. Most typical are formally structured group meetings held at predetermined times. This appears to be the case throughout the organizational hierarchy, from coordinating activities at the "bench" level to communicating and coordinating corporate R&D planning and funding at the strategic level. Frequently, such meetings are designed to purposely include non-R&D people in order to obtain outside perspectives.

An example of the formality of R&D communications found within participatively centralized firms was provided by a chemical-based company which has, for each of its major product lines, semi-annual three-day project reviews. Each meeting is attended by the Vice President for the product line, and staff, including personnel from the technical information center, manufacturing, sales, technical service, and plant production people, as well as anyone else who has a strong interest in the research being discussed. This invitation is typically accepted by a number of researchers from different groups within the firm, and some lively exchanges of opinion, pertinent insights, and experiences occur at these sessions. The meetings begin with presentations to assembled R&D staff by Sales and Production on their particular needs from R&D. The final two days of the meeting are devoted to special interest groups getting together to talk about specific problems. On the last day, a list of projects to be addressed over the next six months is prepared. This list is also open to criticism and debate with the non-R&D attendees at the meeting.

The use of formalized report summaries is another practice widely followed by the participative centralization firms. Each of the firms interviewed made it a practice to issue, on a regular basis, news-letters containing brief abstracts of ongoing research. For the most part, these newsletters were distributed throughout the firm's R&D community and, despite indications to the contrary in the literature on R&D communications, almost all of the managers interviewed believed that these newsletters were valuable techniques for dis-seminating information within the firm, and for whetting appetites to obtain the full reports of studies summarized in the newsletter.

A good summary of the communications found in the participa-tively centralized firms was provided by an R&D executive who observed that his firm's "transfer and coordination approach is very definitely informal and interpersonal, but that it is also backed up by very specific communications such as worldwide engineering standards and common drawing releases. Such formal communica-tions are essential to crystallize the specifics of the technology and agreements. Achieving common standards is no small task and needs strong interpersonal relationships. Also, any attempts to centralize authority can easily result in badly bruised egos as people outside of headquarters feel stripped of authority. Informal communications and good interpersonal relationships can alleviate some of these problems."

World-Market Companies

Two of the firms characterized as world-market firms exhibited participatively centralized R&D management styles. One of these firms was an American corporation, while the other was European. As would be expected with world-market firms, both have a rela-tively high science orientation. In addition, both firms have separated research from development within the corporate structure and have established laboratories in several advanced countries. In both cases, the foreign laboratories of these two companies were established long before the firms acquired a transnational image, and the labora-tories tend to be located in large, important foreign markets where foreign professionals preferred to settle.

Coordination and control of the foreign R&D groups appeared to be similar in the two firms. In the European firm, the directors of the foreign R&D laboratories meet with corporate R&D personnel every three months to coordinate the special tasks assigned to each laboratory. There are coherent research goals in each of the labora-tories, and the tasks assigned to each fit into a total corporate program. The results obtained in any one laboratory become the common property of all. According to the executive interviewed, the foreign laboratories feel equal in status with the central laboratory. There is considerable travel between the laboratories by research managers; at one point, travel was so frequent that the firm employed its own plane on daily flights between England, Germany, France, and Holland. In addition, all research reports are openly exchanged among all of the laboratories.

In the American firm, the foreign R&D laboratories were also located in important markets where the company had established an international presence long before it became a transnational organization. Two of the benefits of such an historical international presence, the firm believes, are the development of a corporate *esprit de corps* that transcends nationalistic feelings and the establishment of an international identity. As a result, there is considerable personal interaction among the R&D management within this firm, and the growth of close personal and family ties are encouraged. The R&D that is pursued in this firm's foreign laboratories is funded by their local companies who, consequently, feel that they have an investment in these units and want to get something out of them. The corporate R&D program is developed by headquarters in collaboration with the foreign laboratories, and the work is formally divided up among them. Thus, within this firm, although the R&D funding for the foreign laboratories is supplied by the foreign affiliates, the R&D programs at these laboratories are established by corporate R&D headquarters.

It is possible to see in both of these companies the influence of a heavily scientific orientation that has been shown earlier to lead to foreign laboratories being regarded in a collegial fashion by central R&D, as well as the strong influence of corporate R&D headquarters that is the hallmark of a participative centralization management style.

COOPERATION

According to DeBodinat, in his original formulation of influence categories,[3] "cooperation" denoted the situation where commitment was reached by agreement between parent and subsidiary approaching equals. In our interviews this agreement among equals best described joint R&D ventures between transnational corporate partners. Within firms, it appeared more accurate to identify the more dominant organizational entity (i.e., corporate R&D headquarters or the divisional laboratory) and to characterize the firm accordingly. As a result, the influence category "cooperation" was not employed in this study.

SUPERVISED FREEDOM

Twenty-three firms were characterized as exhibiting "supervised freedom" in their management of R&D. Thirteen of these were American and ten were European. All of these firms had experience with foreign R&D activities, and 50 percent of these foreign R&D activities had missions that included new product research responsibilities. Firms exhibiting a supervised freedom style of R&D management are likely to have both a number of R&D laboratories abroad and a tendency to place primary responsibility for "operational" decisions in the hands of the foreign R&D management. A central corporate laboratory is also typical of these firms, with funding shared by both the foreign affiliates and headquarters. Coordination appears to be far less formal than that attempted by the participatively centralized firms and tends to rely on good interpersonal relationships and lots of travel. The corporate R&D staffs at headquarters of the supervised freedom firms are typically responsible for reviewing the intentions of the foreign R&D groups and for assuring corporate management that the foreign groups are performing acceptably.

Companies exhibiting a supervised freedom R&D management style appear to use direct placement as foreign entry mode for R&D less than do firms which are participatively centralized, but they appear to use evolution from technical services more than the participatively centralized firms.

Home-Market Companies

Two of the firms employing a supervised freedom approach to R&D management have home-market orientations. Both of these firms are in the nonferrous metals industry, and both are American. In both of these cases considerable freedom is vested in the foreign laboratories yet the firm also has a central corporate R&D laboratory.

In the larger of the two firms, the Director of Corporate Technology is directly responsible for three corporate laboratories and has "dotted-line" responsibility over the operating divisions' labs, which report to him through a corporation-wide technology committee. He is not directly involved with the budgets of the business unit laboratories but does review their plans and receives copies of their program highlights, in an effort to reduce duplication of effort. In general, the Director's involvement with the business unit laboratories is primarily one of providing technical direction. The laboratories themselves are responsible for the monitoring of their projects and programs. While the Director enjoys very close control over the total corporate R&D budget, and thus reviews progress on clearly identified efforts across all labs, his actual direct influence at the business unit level must be conveyed through persuasion or pressure exerted down through the line channels.

The laboratories of the larger firm were established twenty to fifty years ago in an effort to bring the corporation into new technical markets. According to the executive interviewed, there is a great deal of teamwork between the foreign affiliate R&D laboratories and the corporate R&D laboratories. When an affiliate's project is pursued at a corporate laboratory, coordination is effected on a one-to-one basis between a project manager at the corporate laboratory and a project coordinator at the affiliate.

In the case of the smaller of the two firms, there is considerably less R&D undertaken abroad. What R&D is performed abroad is principally tied to the firm's extractive operations and is placed under the responsibility of corporate R&D headquarters, if it is of any significance. This is accomplished through the creation of a liaison person who links headquarters and the field laboratory; often a supporting staff will also be created at headquarters. The liaison managers are an important means of reducing R&D redundancy in this firm. The liaison manager reviews the foreign affiliate's budget after it had been approved by the managing directors of the local subsidiary and the regional group. This review is part of the headquarters's reviewing process, which ultimately includes the corporation's president.

For routine operations, however, the foreign affiliate R&D groups are self-sufficient. If the corporate staff feels a need for innovativeness, it encourages this through persuasion. The foreign affiliates can, however, ignore these suggestions. There is considerable personal interaction between the foreign affiliate laboratories and the corporate headquarters R&D staff in this company. This interaction occurs largely through written communication and travel. The liaison people at corporate headquarters are responsible for ascertaining the technical status of the foreign R&D groups. Although this serves to coordinate technical advancement somewhat, the executive interviewed felt that it is probably more useful in nontechnical commercial activities.

Host-Market Companies

Nineteen of the supervised freedom companies have host-market orientations. Nine of these are American and ten are European. These firms exhibit a wide range of scientific orientations, and most of them have substantial central R&D groups as well as having delegated a significant level of responsibility to their foreign affiliates.

An example of how "supervised freedom" operates as a management style is found in a European tobacco company where the foreign affiliates are profit centers and perform whatever research they need to do for themselves. The corporate R&D group in this case gets involved only through professional critique, although it does review programs from a fiscal and operating standpoint. There is no formal corporate research plan, and the foreign affiliates enjoy considerable independence. The executive interviewed attributes the supervised freedom management style more to the corporation's philosophy than to anything inherent in the nature of the product. He did, however, observe that in industries such as chemicals and steel, where specifications are international, buyers depend upon these international specifications. Tobacco products, on the other hand, are marketed subjectively to local tastes; there are no international standards. Accordingly, the R&D within such firms does not have to satisfy international standards, but rather must meet local needs. Hence, it is possible, if not necessary, to allow the foreign affiliates considerable freedom.

Similarly, in a European chemical company, the R&D budgets of the foreign affiliates are wholly under their own control with no centralized guidance. Each is, in fact, a profit center. There is a research unit at headquarters that attempts to coordinate total corporate R&D efforts through a central committee, but this committee is limited to persuasion rather than direction. The central R&D group does make suggestions as to the introduction of R&D activities abroad, project selection, scope, and procedures, and will occasionally undertake some "exploratory research" for a foreign affiliate.

A diversified European firm, which has limited its foreign R&D activities to the United States, indicated that its foreign laboratories are left free to do what they will, with assistance available from corporate R&D upon request. Any control over the foreign laboratories is accomplished through persuasion and discussion of R&D budgets. While some coordination is attempted wherever it is deemed desirable to facilitate growth in one R&D unit through the cooperation of other R&D units, each unit retains the final say as to the amount of cooperative efforts it commits to. A corporate research office exists in this firm, but without a laboratory and with no formal authority to approve the R&D programs of the foreign affiliate laboratories. Corporate R&D does, however, have a corporate research program for exploratory research that contracts among the three or four affiliate laboratories, which are large enough to have the sophisticated equipment necessary to undertake such work.

In a European food company, supervised freedom is evidenced by the initiation of R&D programs by the foreign affiliates, with corporate management normally accepting them. According to the executive interviewed, "R&D needs to be a self-starting activity, which cannot be built up or cut back quickly through excessive budget changes. There needs to be a complementariness in disciplines, and close ties between research, marketing, and manufacturing, both as to product uses and processes." R&D groups within this firm are located near manufacturing plants for administrative support, with the local manufacturing unit being paid by corporate R&D for this service: "A location near a plant is useful in inculcating a product orientation among the scientists and in facilitating the receptivity of manufacturing to new technical developments. Many of the development groups are even located within a factory."

In some instances, the marketing of products to local subjective tastes is so dominant that the freedom can be nearly total. A food company, for example, has adopted the philosophy that "our foreign managers can run their own business. We expect them to turn a profit. How they do it is pretty much up to them." In this particular firm, all foreign R&D people report directly to the local country manager, with the Vice President for Scientific Affairs having only nominal responsibility for their activities. Foreign R,D,&E budgets, projects, and programs are all determined by the local manager. The entire country budget, however, is determined in the United States, and it is here that the Vice President for Scientific Affairs gets involved, albeit in an advisory capacity. Monitoring of ongoing projects is the responsibility of the local manager. Eventually, when the R,D,&E work is completed, the results are transmitted to the International Operations staff at corporate headquarters and reviewed by the headquarters technical staff. If the technical staff believes that a project has utility elsewhere, they advise the president of International Operations who, in turn, advises the pertinent local managers.

As the previous examples show, firms exhibiting a supervised freedom R&D management style tend to have central R&D groups. An important role of these central R&D groups is to assure top management of the practicality, utility, and soundness of the R&D being performed by the subsidiaries. One R&D executive put it this way: "I serve as a radar antenna for top management, to make sure that our technical programs make sense and are done right. I review plans, projects, expenses, and personnel movement. I have dotted-line authority over each technical director. I review and approve plans. If I say I don't believe in a project they have to defend it, and if we can't reach a compromise at the division general manager level it goes to the group Vice President; if still unresolved it goes directly to the Chaiman of the Board. This latter stage is seldom necessary."

Another executive put it this way: "I am supposed to assure the President that the divisions' capabilities are acceptable and that their technical strategy is sound."

In terms of communications and coordination, the "supervised freedom" firms tend to rely on informal relationships between R&D laboratories for coordination, with far less use of routine formalized meetings among laboratory personnel than that found among the participative centralization firms. The supervised freedom firms do, however, sponsor considerable travel by individuals on an ad hoc basis for coordination purposes. In several instances, these informal coordination practices have had dysfunctional consequences. In one chemical-based firm, for example, the first steps are presently being taken to move to a more centralized R&D management system: "Our first working relationship has been established with our British facility. A development program has been identified which the British group is interested in, and while the corporate lab in the United States is funding this work the actual R&D is being performed in the United Kingdom and managed by the U.K. research director. We selected the British lab for this effort because they had not been performing at an acceptable level under the previous loose coordination." In addition, the firm's Group Vice President for European Operations has recently agreed to allow the group's R&D plan to be reviewed by the corporate R&D laboratory for comments. There will be, however, no transferral of line responsibility.

A second example involves an American food firm that is presently initiating a central European R&D laboratory in response to the overseas subsidiaries' expressions of need to become more involved in new product development and to obtain more responsive product and process assistance than was possible under the existing informally coordinated system.

Although most of the supervised freedom firms acknowledged the possibility of duplication of efforts resulting from informal coordination, none of them appeared to be seriously troubled by such possibilities. In most cases they felt that the strong interpersonal relationships that existed among the laboratory directors would be sufficient to uncover duplication before it grew to serious proportions. An R&D executive in a chemical-based firm stated that: "In every single year that I've held this job, I've had one division help a sister division technically. During our project reviews I learn about what the divisions are doing and I act as a clearinghouse. Presently, a special process invented by one of our divisions is being used by another division in a totally different business as a result of my informal suggestions. This clearinghouse role has avoided costly reinnovation."

Several of the supervised freedom firms employ a lead-division format for the coordination of their R&D. In the case of a pharmaceutical company, most, but not all, of the lead divisions are located in the United States and are responsible for support and granting of clearance for new technical advances in particular product areas. An energy-based company has lead divisions which are supposed to supply the technology directly to the overseas subsidiaries. Coordination of day-to-day technical activities occurs directly between the lead divisions and the overseas operations, without the involvement of the corporate technology group, and accounts for about 70 percent of the total R&D coordinating effort. If, however, new processes or new materials are involved, then the corporate technical group does join the process.

The importance of a central R&D facility to firms exhibiting a supervised freedom management style is particularly striking. Almost every supervised freedom firm had a central R&D lab. A major function of these laboratories is to perform long-range research that would typically not be pursued by the operating divisions. In one machine tool company, corporate R&D focuses entirely on new ventures rather than extending existing product lines. Decisions on new ventures are made by the corporation's President, Executive Vice President, and the Vice President of R&D. In a chemical-based firm, the central R&D group works only on projects with a common interest to the operating divisions, yet which none of the operating divisions would pursue independently, because of prohibitive costs.

The mechanisms for funding the central R&D in the supervised freedom firms are as varied as the firms themselves. In most cases, however, it is a mixture of corporate seed money and allocations assessed against the foreign affiliates. In several instances, the central lab also performs contract R&D for external clients. One interesting perspective on the funding of corporate R&D was made by an executive in a machine tool firm who observed that their central laboratory was formerly funded by a prorating scheme among the divisions, until it was realized that such a scheme was "allowing middle management to determine the future of the company."

The reliance which the supervised freedom firms placed upon their central R&D laboratories, combined with the heterogeneity that exists among their affiliate laboratories as a result of their relative freedom, appears to lead to a selective restriction of scientific and technical communications within these firms. Communication networks that are indirect, by virtue of having to pass through a central laboratory, present the opportunity for selective interference, whether intentionally or unintentionally. A tobacco company executive observed, in this regard, that the major R&D groups within his corporation interact with each other on a one-to-one basis. Smaller subsidiaries, however, without technical groups have to rely upon the central corporate laboratory to link them into the interlaboratory communications network. As a result, the central corporate laboratory is in the position of deciding what information is passed on to these smaller subsidiaries.

An example of the confused communications situations that can develop in supervised freedom firms with host-market orientations, was provided by a machine tool company that encourages its foreign affiliates to communicate freely among themselves, but selectively disseminates any information that is channeled through the central R&D group. The foreign affiliates have taken advantage of free bilateral communication to the extent that one division has one of its R&D people semipermanently employed, full-time, at another foreign division's laboratory, in an effort to influence process equipment development. At the same time, however, any information passing through the central laboratory is shared with the foreign affiliates on a need-to-know basis, determined by the Vice President for Corporate R&D.

Supervised freedom can also encourage laxity in scientific and technical communications. In one instance, a firm characterized as employing a supervised freedom management style required that all corporate technical operations issue annual reports, yet the German R&D group had only issued six reports in the previous ten years, all of them within the past two years. Furthermore, until recently, there had been no visitations to the German laboratory by U.S. headquarters personnel in five years.

World-Market Companies

Two of the firms exhibiting supervised freedom management styles have world-market orientations. Both are large American electronics-related companies. Both firms have a research laboratory in Europe and one also has a research laboratory in Japan. In all cases, the foreign research laboratories are performing highly sophisticated, fundamental research and are relatively self-sufficient within a specific technical area. The laboratories were all originally established to tap existing foreign scientific and technical communities. Coordination and communication among the laboratories in both firms are marked by considerable travel and informal contacts between the directors of the laboratories.

The major difference between the two companies is in the relationship between the central R&D laboratory and the operating divisions. In one of the firms, virtually the entire research program is determined by the central research laboratory itself. Its budget comes from the corporation in the form of a grant, although the proposal for funding that Research submits is criticized by the operating divisions. In this particular firm, however, Research feels very strongly about its charter to anticipate future problems, products, and needs, and consequently, does not entertain many project suggestions from the operating divisions.

In the second firm the budget for Research can be broken down as follows: 60 percent for activities requested by the operating divisions, 20 percent for activities that Research believes will support the operating divisions, but that originate within Research and that might not even be recognized or acknowledged by the divisions, and 20 percent for exploratory research. Projects in the first category are set against specific business objectives of the divisions. Working task forces consisting of key people from Research and from the division's engineering group develop the projects. These task forces are retained during the year to review the projects quarterly and often monthly. Projects in the second category are reviewed simultaneously with those in the first category; during the past two years, almost all of the category two projects have been accepted by operating divisions. Projects in category three are chosen by Research. These choices are communicated to the rest of the corporation via a top-level business development committee. Marketing and Product Planning are integrated into these activities as necessary and appropriate. The business development committee is chaired, however, by the Executive Vice President for Research and Engineering, and this committee has the authority to recommend new organizational structures to the corporation's president in order to enter new markets.

The supervised freedom management style employed by this second firm is most evident, however, in the amount of R&D performed at the foreign affiliates. Each division is responsible for its own worldwide technical activities, and they operate just like any other comparably sized competitor. The Corporate Vice President of Research and Engineering is charged with assuring the Corporate President that each division's technical capabilities are acceptable and that their technical strategy is sound. His actual line responsibility, however, is limited to the work that is done in the corporate laboratories in support of the divisions.

TOTAL FREEDOM

Only three firms in our sample could be characterized as exhibiting total freedom in their relationships with their foreign R&D groups. All three companies had established their foreign R&D efforts through acquisition, and in two of the cases the skills and interests of the foreign laboratories were considerably different than those of the central corporate laboratory. In these latter two cases, the foreign R&D effort was, at this point, clearly an anomaly. Their inclusion in this study is solely for completeness and to present the difficulties associated with the management of transitional situations.

Home-Market Companies

Two of three total freedom firms have a home-market orientation. In both cases the foreign R&D effort is clearly different from the dominant thrust of the parent company. Both firms appear to be ignoring the foreign R&D effort by maintaining its funding but leaving it totally on its own. In one of the cases this is the result of a recent acquisition, and there were suggestions that this lack of involvement would not last forever. In fact, discussions on technical communications had already begun, but the research manager had yet to determine how this relationship would develop. According to him, there was little incentive for involvement because there were no really important technical reasons to share research between the two organizations.

In the case of the second firm, a small European lab of three or four people has existed for some time following the acquisition of its parent firm. "These people continue to make and sell their own products just as before the acquisition and to sell our products which we make for the U.S. market. In addition, we take what they make and sell it in the United States, as is, without any modification." This total freedom relationship has existed for several years and the foreign affiliate is, at present, the parent firm's only foreign R&D activity. Apparently, total freedom in this case is an artifact of a small, inconspicuous, foreign R&D profile. Several years earlier, however, the firm had a small metallurgical applications and process group of ten people located in Sweden. Headquarters decided to disband the group because it was too small. There was no public reaction to the decision, but the parent firm had to pay each worker one year's separation pay.

Host-Market Companies

There is only one host-market company exhibiting a total freedom management style with regard to its foreign R&D activities. This particular firm has experienced a rather rapid expansion of its activities in foreign markets through acquisitions. According to the Director of Research of this chemical-based firm, "Any time we want to get into something new, we buy someone. Its quicker and safer! The firms acquired, however, tend to be run by strong entrepreneurs who resent central controls. We pride ourselves on being unobtrusive. As a result, however, there is an extremely low level of research interaction between the domestic divisions, the technical center, and the foreign divisions." As an indication of how minimal the research interaction is within this firm, although all of its overseas subsidiaries are serving the same product market as the parent firm, the Director of Research has been unable to convince the overseas R&D groups to even register their projects with the central lab to reduce duplication. Furthermore, the Director of Corporate Research has not been to many of the overseas R&D laboratories, including the Brazilian lab, which is the firm's second largest overseas operation.

World-Market Companies

There were no world-market companies that exhibited a total freedom management style in their relationships with their foreign R&D affiliates.

SCIENCE ORIENTATION

Throughout this study, the scientific orientation of the firms has been linked to various behaviors. In order to examine the relationship between science-orientation and managerial style, it was decided to look at only the host-market companies, to avoid the small sample problems presented by the other market-orientation categories. Furthermore, the total freedom firms were omitted because of the transitional nature of this particular management style, and the absolute centralization firms were omitted because of the unsuitability of that management style for foreign R&D activities.

The data obtained from the interviews indicate that those host-market firms employing a participative centralization management style typically have a strong science-orientation—namely, firms in high-technology industries such as electronics, pharmaceuticals, and chemicals. When one considers the numerous products such firms sell, and the markets they sell in, the need for centralization becomes clearer. As one R&D executive put it, "If your research is related to products, and you have a myriad of products and markets, you need a high degree of coordination. When you're dealing with one or three metals, you're probably familiar with everything going on, and you can enjoy more informal coordinating schemes." Science-orientation, then, appears to be a useful explanation for the choice of a firm's R&D management style.

SUMMARY[4]

The strong similarities among the American and European transnationals examined suggests that choice of coordination and control systems are based on considerably more than geographic or cultural proximity.

A firm's management style, however, does appear to be partially determined by market-orientation. Two of the four managerial styles, absolute centralization and total freedom, appeared in less than 20 percent of the firms and seemed to be anomalies in terms of understanding how management style affects the coordination and control of foreign R&D activities.

Almost one-third of the firms, however, had adopted some form of participative centralization as a management style in relating to their foreign R&D activities. These firms tend to have a large, active R&D program in foreign locations and to be quite willing to assign new product research missions to these foreign laboratories. They exert strong centralized authority over the funding, program, and project selection decisions of the foreign R&D groups, and they tend to employ a structured, often sophisticated, coordination and control system. Participative centralization firms appear to have a strong science-orientation and to rely almost equally upon direct placement and evolution from technical services as entry modes for foreign R&D actvities.

Nearly one-half of the firms interviewed employed a supervised freedom management style. These firms also tended to pursue new product research in their foreign R&D groups but also to attach great importance to a central corporate R&D laboratory. This central laboratory's function was usually to review the intentions of the foreign R&D groups and to assure corporate management that these foreign R&D groups were performing adequately. Primary responsibility for the activities of the foreign R&D group resides, however, with the management of the foreign affiliates. Coordination of R&D activities appears to be far less formal among the supervised freedom firms than among the participative centralization companies, as the former tend to rely more on good interpersonal relationships and extensive travel. The supervised freedom firms appear to favor evolution from technical services as an entry mode for foreign R&D rather than direct placement.

The scientific orientation of a firm also appears to significantly influence the coordination and control system developed by transnational corporations to manage their foreign R&D activities. Firms with strong science-orientations tend to employ participatively centralized management styles.

NOTES

1. Henri DeBodinat, *Influence in the Multinational Corporation: The Case of Manufacturing.* Unpublished doctoral thesis, Graduate School of Business Administration, Harvard University, 1975.
2. Using the OECD Science-Base of Industries Continuum. This continuum reflects the relative degree of an industry's research intensiveness, in the United States in 1962. To qualify as science-based, an industry must have R&D expenditures amounting to 2% or more of sales, or scientists and engineers amounting to 1% or more of total employment. For a discussion of this scheme see J. E. S. Parker, *The Economics of Innovation*, (London: Longman, 1974), pp. 146-149.
3. DeBodinat, *ibid.*
4. An empirical analysis of the information reported on in this chapter is found in W.A. Fischer and J.N. Behrman, "The Coordination of Foreign R&D Activities by Transnational Corporations," *Journal of International Business Studies*, Winter 1979.

Chapter 3

Invention and Innovation in Foreign R&D Activities

The case studies that were undertaken as part of this study and a considerable amount of research (none of it specific to transnational corporations) have established the very important role that managerial style, coordination, and control play in influencing patterns of invention and innovation within R&D groups. Since invention and innovation are the reasons for establishing R&D activities, anything that affects their achievement is of interest to policymakers. The previous two chapters have reported the relationships among market orientation, managerial style, and the foreign R&D location decision. This chapter extends that examination to include the relation of these aspects to the patterns of invention and innovation abroad.

PATTERNS OF INVENTION AND INNOVATION ABROAD

Patterns of invention and innovation abroad are significantly .affected by the market orientation of the parent and affiliate and by the R&D management style, which permit the foreign lab more or less autonomy.

Home-Market Companies

By their very nature, the R&D efforts of the home-market firms are dominated by their domestic R&D activities. To the extent that they are involved in foreign markets, the home-market firms tend to view these markets as extensions of their domestic markets, and do not undertake R&D especially for them. Since they are often dealing with bulk, or commodity-type, products, this avoidance of R&D for particular foreign markets is understandable. In one American metals firm, the Technical Director perceived foreign technical activities as: "largely applying what's been done in the United States. Most of the time we're only doing R&D for the U.S. market. There is no case of pursuit of a 'foreign' market. The 'foreign' market is essentially fallout. We do very little modification for foreign sales." Accordingly, in this firm, R&D personnel do not enjoy much foreign exposure or travel. The R&D group gets its best information about foreign needs through sales and marketing visits to customers. This company's foreign marketing group, however, is technically trained, which makes it a particularly useful information channel.

A second metals-based firm, with a small R&D group in Latin America, reported that "international operations are not as prone to technical change as are domestic operations. There are different motives for employment and efficiency abroad and there is often no competition to spur motivation. Except for a moral responsibility to maintain a good working environment, and to protect the physical environment, there is less reason to upgrade foreign operating facilities as much as is done in high labor-cost markets. Accordingly, major technical operating changes are made when you change units or when incremental growth occurs. At each of these investment decision points, the firm's existing R&D resources are supplemented by using outside engineering consultants. In markets where labor costs are high and labor limited, such as Australia, for example, a more continuous updating of facilities is mandated in order to reduce operating costs. In some other cases, innovation is determined by local energy availability, or by environmental problems."

Most of the initiative for new projects in this second firm comes out of a three-person technical group at corporate headquarters. This corporate headquarters group acts as a guiding force, assisting the foreign R&D group in putting projects into an acceptable time and budget framework. The headquarters group also assigns someone to a liaison position for control purposes. Small projects are run at the overseas laboratory under the responsibility of local people. The local R&D managers are, in turn, responsible to local corporate management who, in their turn, are responsible to a liaison manager at corporate headquarters. If a project becomes too big, a managing director for the project is attached to the corporate group at headquarters, and a special staff for the project is created at headquarters.

In both of these firms the focus of R&D within the corporation is on the domestic market, and R&D for foreign markets, to the extent that it is done at all, plays a minor role. This is most evident in the second firm, where, in the event that an important project does come about as a result of foreign operations, it is quickly brought into the headquarters group and controlled there.

Host-Market Companies

The patterns of invention and innovation found among the host-market companies reflected not only their market orientation but also their choice of managerial style. Since this market grouping has enough firms to allow a comparison of the affects of managerial style on invention and innovation, the discussion will be subdivided to look at both the participative centralization and supervised freedom firms.

Participative Centralization Companies. Those host-market firms exhibiting a participative centralization management style employ rather complex procedures that must be followed in order to obtain approval to initiate a project; corporate R&D headquarters also has a marked tendency to assign work to laboratories and balance overall corporate scientific and technical capacity among domestic and foreign laboratories. In a number of these cases, core research is largely performed at corporate headquarters.

An example of the innovation process within such a firm was found in a chemical company. In this firm, a project requiring basic research must be done at corporate headquarters, where such resources exist. The bench researcher who originates the idea must sell it to the local manager, who, in turn, submits it to the geographic area president. From there, the idea is passed on to R&D Administration at corporate headquarters for approval. According to the executive interviewed, R&D will, in general, go out of its way to accommodate such requests, however: "R&D Administration does represent 51 percent of the final vote." Within this same firm, technology centers for each operating division exist at corporate headquarters. Their job is to ensure that each of the firm's labs is utilizing the most up-to-date technology. Personnel at the technology centers travel frequently and audit the foreign laboratories. Competition involving contributions toward productivity, cost reductions, Btu reductions, and the like are held on a company-wide basis and the winning ideas are disseminated throughout the firm. In addition, once every two years, all of the plant managers producing a certain product will meet to discuss their experiences. According to the executive interviewed, the firm's experience with this sort of coordination has convinced this company "that we do not need to do process research all over the world. Instead, we have committed our foreign R&D resources to product-applications research." In support of this philosophy this firm employs a global product coordinator scheme to centralize coordination and control over all research being performed on a particular product. The company believes that these coordinators, the technology center audits, and marketing are the predominant sources of R&D ideas for their overseas R&D groups.

The complex process of initiating R&D projects in participatively centralized firms is further illustrated by a pharmaceutical firm where projects initiated in the foreign laboratories must be signed off by both the regional marketing supervisor and the regional R&D supervisor before being submitted to corporate R&D headquarters for approval. In a second pharmaceutical firm, candidate projects enter into the R&D decision process through a Project Development Committee composed of general managers from all areas, such as marketing, sales, and manufacturing. Updating is performed by a Commercial Development Research Committee in consultation with all overseas R&D units. This group determines when it is appropriate to tool up, to register a new product, and when to market a new product. Personnel involved in foreign regulatory affairs report to this committee. Communications in this company includes a summation of monthly highlights of bench reports that is sent to all laboratories. They are expected to critique these reports and to communicate these critiques directly to the section heads responsible for the reported work.

The centralized nature of participatively centralized firms allows them to consider total corporate R&D resources and to assign projects to laboratories in a way that will balance work with available capabilities on a corporation-wide basis. This can be seen in a pharmaceutical company where new ideas are assigned to a laboratory based on its available capacity, interests, and skills. This permits the firm to shift workloads between foreign and domestic laboratories.

The ability of such participatively centralized firms to balance their R&D capacity can also be seen in a machine tool company in which an idea originating in Britain was transferred to the United States for engineering and final development work by temporarily transferring the entire originating British team. The transfer was effective because of the historically high performance of the U.S. group, a current work overload at the British lab, and the technical director's belief that the U.S. group needed some morale boosting.

Another pharmaceutical firm employs five- to twenty-five-person research committees at both its foreign laboratory and its domestic laboratory. These committees apply to the corporate research director for approval to do clinical work on particular compounds. If approved, the headquarters staff establishes project teams of four to five people, including representatives of process design, toxicology, formulation, and clinical research, which have responsibility to do everything necessary to get the compound to the market. Each representative on the team links it to his or her research area; for example, the clinical representative would also report to the Clinical Research Planning Committee in order to coordinate clinical research in support of the team's goals. Every sixty days the team reports to headquarters on its progress.

A machinery-based, participatively centralized firm also employs teams to coordinate corporation-wide resources in the innovation process. Once a major development arises, mini-conferences of bench researchers, pulled from the worldwide corporate R&D resources, are held at a working level. In addition, small groups are sent around the world to visit plants and markets in order to obtain information essential to a successful innovation.

Supervised Freedom Companies. In general, the host-market firms exhibiting a supervised freedom managerial style tend to apply a simpler decision process regarding innovation than do the participatively-centralized firms because of the greater autonomy enjoyed by the foreign affiliates. In one food company, for example, where the foreign affiliates have almost total responsibility for project selection and, indeed, for whether or not they should even do R&D, the corporate headquarters R&D staff gets involved with innovations only to the extent of making recommendations regarding the product when the local manager applies to corporate International Operations for permission to market the product—after the R&D has already been performed! The R&D executive interviewed offered this scenario to illustrate the freedom enjoyed by this firm's foreign affiliates: "If an operations manager in a foreign affiliate wanted to market a product, and we [the corporate R&D staff] knew something about the dietary conditions in that particular country and thought a particular additive would be desirable, we would suggest that it be put into the product when the local manager applied for permission from International Operations to market the product. Although it has never happened, the local manager could resist such advice, but he would have to offer strong reasons for such action." In terms of the intracorporate transfer of innovation, this same executive remarked that "It's up to the foreign manager to call upon whomever he wishes to obtain new technology. If it's a product, he'd most likely go to the Senior Vice President for Corporate Development for such advice." If and when called upon, the Corporate Development staff could then suggest ideas developed elsewhere within the firm that might be appropriate. A task force, combining Corporate Development staff personnel and representatives of the potential recipient of the innovation would be appointed by Corporate Development and charged with adapting the idea to the recipient's situation. The cost of this effort would be borne by the recipient. Here we have a typical case of supervised freedom: the impetus comes from the field, the costs are borne by the field, and a central corporate lab, normally devoted to long-range basic R&D, supplies assistance, when asked, on effecting a transferral of R&D capabilities.

In one chemical-based firm, the situation is somewhat similar and serves to reinforce the conceptualization of supervised freedom. The products of this firm are not highly standardized, but they are similar within a broad framework. There are basically three or four product types, which differ slightly by chemical compositions and applications. To varying degrees, these product differences are the result of market differences, tradition, different manufacturing equipment, and to some extent the "not-invented-here" syndrome (a well-documented negative reaction to ideas imported from outside the group). It is therefore difficult to determine the actual degree of standardization among the firm's products.

Recently, the firm's U.S. division developed some new, superior products and wanted to exploit them worldwide. This task was given to the central R&D group, who created two international development teams comprising technical representatives from the United States, the United Kingdom, and Germany. These teams were charged with determining how these products could be produced in different countries and with standardizing the products. This has been going on for a couple of years, and the firm has achieved some degree of success within the European Economic Community and the United States.

The supervised freedom firms tend to highlight central R&D activities in their discussion of innovation. For the most part, these central laboratories work on projects detached from the efforts of the operating divisions laboratories in terms of technical focus, time frame, or both. In one machine tool company, each product group has a New Product Planning Committee, which consists of representatives of Research, Marketing, Engineering, and Market Research, and which reports to the Group Vice President. These New Product Planning Committees meet on a regular basis to develop product strategy and to allocate the group's R&D resources. Products within a group are the group's responsibility, and each group funds its own R&D. Corporate R&D focuses on funding new ventures rather than extending existing funds of the corporation. New venture decisions are made by a Commercial Development Group within central R&D. This group emulates, for the central lab, what the New Product Planning Committee does for an existing product group. The Commercial Development Group is sensitive to such factors as markets, competition, and patents. It provides the intelligence necessary for new market decisions. The executive interviewed felt that "Marketing is not useful in such decisions because these decisions involve totally new ventures which Marketing is not familiar with." Products within the machine tool industry are typically not standardized. Traditionally, modifications are made over time and on the basis of the customers' particular needs. Machine tools are generally manufactured in a job-shop environment; an eventual production volume of fifty for any one machine is unusual. With the closing of the U.S.-European technology gap, however, this firm is beginning to anticipate adaptation for foreign markets at the *onset* of the design process rather than at the end of the domestic product life cycle. The New Product Planning Committees pinpoint the potential markets and decide whether or not to standardize from the start. In one recent case, the decision was to go after the domestic market while fully expecting to adapt later if the product proved successful. A second recent product, on the other hand, has been in development during the past eighteen months; while it was still in the design phase, the firm began developing a factory option for the foreign market. This latter development is being undertaken by corporate R&D groups belonging to its foreign affiliates.

One chemical-based firm, however, that exhibits a supervised freedom management style has become disillusioned with the role that a central research laboratory plays. The executive interviewed stated: "I abolished our central research lab a few years ago because nothing came out of it. It was a beautiful, isolated campus of people busy on things not commercially worthwhile." Presently, this firm allocates corporate seed money to divisional laboratories for projects which would be difficult to prove commercially important at present, but which are thought to be promising. These seed projects are performed in the group laboratories and are expected to show commercial volume within three years.

World-Market Companies

Since world-market firms stress standardization of their products, the innovative process is tightly coordinated. However, the community of peers, which also tends to characterize their corporate R&D resources, results in more of a spirit of negotiation among equals than it does in other firms. It is only the essential authority residing in and exercised by the central R&D headquarters in these firms that prevents them from being described by a "cooperative" influence style. An R&D executive in a world-market firm explained that projects in his firm typically originate as a result of joint planning between the firm's domestic and foreign laboratories. The R&D in this company's foreign affiliates is funded by the affiliates, but the programs and portfolios of the foreign laboratories are considered to be integral parts of the firm's worldwide R&D strategy. This strategy is formulated at semiannual meetings of the laboratory directors where they review activities and discuss shifts in the allocation of the firm's R&D resources. These discussions also lead to the development of a list of corporate priorities, which is then formally divided among the laboratories.

Coordination between the laboratories in a second firm is conducted on an informal basis between the lab directors. It is quite likely that one lab will seek permission from the R&D director to pursue a project that is being simultaneously pursued by a sister laboratory. In other instances where sister laboratories are jointly working on a project, a coordinator is appointed whose role is to complain to the R&D director if one of the laboratories is not performing adequately. In such cases, it is the R&D director's responsibility to speak with the director of this laboratory and negotiate a solution. Much of this interlaboratory communication is carried out at the project level, often by telephone.

Research projects in this firm originate almost entirely in the Research group. There is not much direct input from Marketing, a result of the company's having assigned to the central laboratory the task of pursuing very advanced ideas, while the development of current products is performed within the appropriate business division. The central lab's mission thus becomes understanding the firm's problems and contributing to their solution and to scientific advancement. As a consequence, there is no list of desired projects presented to the Research Division, and according to the executive interviewed, those projects that are suggested externally typically turn out to be "relatively unimportant and/or utterly impossible."

STANDARDIZATION OF PRODUCTS AND PROCESSES

In most of the firms interviewed, including the host-market companies, there was a continuing effort to achieve a greater degree of process and product standardization. Strong support was shown in almost all of the firms for products which could be offered on a worldwide or at the very least a regional basis. Several firms noted that they would no longer pursue R&D on a product unless it could be sold in a number of national markets without modification. Furthermore, no firm was willing to adapt production processes to reflect host-country resource endowments, other than in trivial respects peripheral to the core technology. (The employment of human carriers rather than fork lift trucks was a commonly mentioned process adaptation undertaken in countries with high labor availability.) Where firms did adjust their production processes for foreign operations, it reflected production volume (i.e., market size) rather than host-country factor endowments.

The implications of these attitudes are quite significant for developing countries. For one thing, they suggest that transnational corporations should not be counted on as developers and carriers of so-called "appropriate" technology. Furthermore, the very attributes that make a developing country market attractive to a transnational corporation as a site for locating R&D (i.e., market volume) are also the attributes that discourage the employment of so-called "appropriate" production technologies, because of the propensity of these firms to employ sophisticated manufacturing technology to meet large market volumes.

FACTORS AFFECTING THE SIZE OF FOREIGN R&D GROUPS

The ability of foreign R&D groups to fulfill their mission to invent, to innovate, and to evolve into missions of increased responsibility depends upon a number of factors. Some of these factors, such as methods of coordination and control, communications, management style, and the present firm's objectives regarding product and process standardization, have already been addressed in this chapter. Another factor that was found to be particularly important in our interviews was the "size" of the R&D group. In many interviews, the ability to achieve a certain minimum size in an R&D group was among the primary considerations in determining whether or not to do R&D abroad. In a smaller number of interviews, mention was made of foreign laboratories that have been closed down because they failed to contribute to the parent firm's needs. In these cases, the lack of sufficient size was often cited as a prime contributor to this failure. The term typically employed to describe the size of an R&D group necessary for inventive and innovation performance was "critical mass."

The size of any particular R&D group is a function of many variables. Still, R&D managers consider that an R&D group can be too large or too small, leading to the concept of critical mass. The critical mass implies some threshold size for an R&D group that is necessary for the achievement of a level of output that justifies the group's continued existence. The critical mass for an R&D group is a function of the group's mission, the nature of the science and technology involved, and the abilities of the professional and support personnel. By and large, it is that size necessary to ensure rich communications both within the group and between the group and its environment, to allow the degree of scientific and technical differentiation among the group's personnel necessary for fulfillment of its mission, and to acquire whatever instrumentation and organizational slack necessary for acceptable performance.

The concept of a critical mass relates also to R&D productivity; it is a recognition of the economies of scale that affect the performance of R&D. These were shown earlier to be a principal reason why transnational corporations are not more active in locating R&D units overseas. Since R&D is essentially an information-creating and -sharing activity, it is essential that it be done in groups. Furthermore, heterogeneity of projects and people is particularly important for product-oriented R&D productivity. Thus, to be successful, an R&D unit needs groups of groups. In addition, the modern instrumentation and facilities necessary for product-oriented R&D are expensive to the point where, with a small group of researchers, the fixed costs of establishing a facility will overwhelm any reduced variable costs of professional salaries in the developing world. Thus, the feasibility and nature of overseas R&D activities of transnational corporations is affected by an industry's critical mass for R&D, and the ability of the local scientific and technical infrastructure to support these needs.

Response of R&D managers varied considerably as to what the critical mass for an R&D group would be in any particular industry. A number of firms reported closing foreign laboratories because they were too small to be productive. In most cases, these were firms in high- and medium-science industries, and the laboratories consisted of less than ten technical people. Industries serving consumer markets could reach a critical mass with a smaller R&D staff than can highly scientific industries. Furthermore, R&D groups in consumer-oriented industries appear, in general, to require less sophisticated personnel and less variety in personnel specialization than do R&D groups in the science-based industries. Table 3.1 provides some evidence supporting these observations.

SUMMARY

Differences in market-orientation appear to exert a significant influence on the patterns of invention and innovation in a firm's foreign R&D activities. In general, companies with a home market orientation appear to treat whatever small foreign R&D effort they support as nothing more than an appendage to their domestic R&D activities. Host market firms, on the other hand, are much more deeply involved with their foreign R&D activities, although the degree and nature of this involvement depends upon the management style adopted by the firm.

Considerable differences among host market companies were found as a result of their management style. The participative centralization firms are much more tightly coordinated with the central R&D group closely involved in project monitoring and paying considerable attention to detail. As a result, these firms appear able and willing to allocate the firm's R&D resources on a global scale. Supervised freedom firms, on the other hand, although also maintaining central corporate R&D groups, are far less intimately involved with the R&D efforts of their foreign affiliates. Their central R&D efforts often specialize in long-range or new business activities that isolate them from many of the foreign affiliate laboratories.

World market firms tend to have tightly coordinated international R&D activities, largely as a result of the worldwide standardization of their products and processes. Their reliance on sophisticated foreign scientific and technical talent and the fellowship that

Table 3.1. Estimates of Critical Mass for R&D

Industry	Critical Mass Estimate	Remarks by Company Executives
Pharmaceuticals	100-200	To obtain specialty differentiation
Pharmaceuticals	60-80 technical people	8-9 Ph.D.'s in any one area
Pharmaceuticals	50 technical people	Better off with 100-400
Chemicals	25-30 technical people	Number needed to afford instrumentation and to ensure interaction
Paints and chemicals	25-30 technical people	Necessary for interaction
Paints and chemicals	2-5 technical people	Short-range technical projects only
Transportation components	200-300 technical people	This type of applied R&D requires familiarity with many different technical areas
Automotive	20 technical people	Applications R&D
Components	150 technical people	Fundamental R&D
Machinery	6 technical people	
Tobacco	5-6 technical people	Sufficient, not broad research capability
Food	1 technical person	Plus some subcontracting

Source: Interviews presented in W.A. Fischer, *Institutional Development of Appropriate Industrial Technology in Developing Countries: R&D Policies and Programmes.* Background Paper for the Working Group on Conceptual and Policy Framework for Appropriate Industrial Technology, United Nations Industrial Development Organization International Forum on Appropriate Industrial Technology, New Delhi/Anand, India, 20-30 November 1978, ID/WG.282/90.

such talent inspires leads to management by negotiation rather than by direction. Accordingly, a large amount of decisionmaking and R&D resource allocation in such firms is accomplished through informal negotiation between laboratory directors.

Standardization of production processes and products appears to be a fairly common objective among most of the companies interviewed. While there was some willingness to modify a production system to build in more labor-intensive activities at the periphery (materials handling, for one, often mentioned), no firm was willing to make a major change in its production processes to accommodate different foreign resource factor endowments. To the extent that significant adaptations to process machinery or design are made, they are typically a function of market size, rather than local resource endowments.

The idea of a "critical mass," or minimum effective size for an R&D group was subscribed to by most of the managers interviewed. Although there was considerable variation in the estimates of critical mass, it is clear that it varies with the scientific orientation of an industry and the type of R&D to be performed. Generally, consumer-oriented industries and process-oriented R&D require the least critical mass. This can have significant effects on both foreign R&D location decisions and mission assignments if a transnational corporation perceives that a particular foreign affiliate will be unable to staff a laboratory at a level sufficient to reach and maintain critical mass. In general, the concept of critical mass suggests that if R&D activities are to be established in developing countries by transnational corporations, they most likely will be low-technology, development-oriented activities.

Chapter 4

Collaborative International R&D Activities

Collaborative commercial arrangements between parties of differing nationalities, in the form of joint manufacturing ventures, have long been recognized as a useful means of establishing capital investment, transferring technology, and developing local commercial infrastructures. Although the performance of collaborative R&D arrangements has not received similar attention in discussions of international trade and development, quite a few of the firms interviewed in this study were engaged in some form of collaborative R&D venture. In most cases the collaborative arrangement took the form of a joint manufacturing venture, often in a third country, that had R&D attached to it or required R&D support from the parent firms. Some instances of research-sharing arrangements unaccompanied by institutional partnerships, such as joint R&D ventures and cooperative R&D projects, were also found.

FOREIGN MANUFACTURING JOINT VENTURES REQUIRING R&D SUPPORT

At least twenty-eight specific *manufacturing* joint ventures between parent companies of different nationality that required

R&D support of some form surfaced in the interviews. Both American and European firms were involved in these joint ventures and tended to favor developed countries when such activities required R&D, although some use was made of small countries among the American firms that was not evident in the European sample (Table 4.1).

Most firms stated their willingness to accept the joint venture as a way of life in 1978, and many were willing to accept minority ownership positions if it became necessary to gain access to a market. In those joint ventures where R&D and/or technology sharing is of some importance (and not all joint ventures are characterized as such), the degree of participation in the joint venture's R&D appears to be a direct result of the ownership position of the transnational partner. In slightly more than one-half of the joint ventures, the transnational corporate partner regarded the joint venture as simply an extension of manufacturing and marketing opportunities to exploit innovations, or as an "add-on," rather than an effort necessitating the altering or expanding of the corporation's R&D program. In every case but one where the actual ownership position within the joint venture was ascertained, those firms that held a minority (or, in two cases, 50 percent) ownership interest had adopted the philosophy that the joint venture was simply an "add-on." Furthermore, their description of their R&D plans for these joint ventures clearly indicated a reluctance to become involved with

Table 4.1. Locations of Foreign Joint Ventures Requiring R&D Support

As Reported by American Transnational Corporations		As Reported by European Transnational Corporations	
Australia	3	India	5
Japan	3	United States	3
Brazil	2	France	2
Holland	2	Germany	1
Denmark	1	Japan	1
India	1	Spain	1
Mexico	1		
Peru	1		
Spain	1		

Source: Interviews.

supporting R&D for a corporation they did not control. This was summed up concisely by an European firm faced with a host-government-imposed joint venture in a developing country: "Given the coming joint venture, the laboratory there will be gradually disassociated from divisional support so that there is no continuing flow of information when we assume a minority position. Our staff professionals will simply not see it as advantageous to make a visit to such a laboratory, and it will gradually be cut off. And, since minority joint ventures are not under divisional control, contacts are greatly reduced to the point where the venture is treated almost as a portfolio investment."

In other cases where firms held a minority position, R&D activity in the joint venture occurred without support by, or access to, the minority parent's R&D activities. As an example, another European transnational, forced by a host-government into establishing an R&D activity in a developing nation, indicated that the laboratory was actually not accomplishing very much but was there simply to satisfy the host government. In the second case, in the same developing country, the transnational firm interviewed had indeed established an R&D facility, but all of the funding for the lab's R&D program came from the foreign subsidiary, where there is little understanding of the role of R&D, and where any initiative for R&D support must come from the foreign subsidiary. Several R&D managers stated unequivocally that, as long as they did not have control over the joint venture, it would be held at arms length from the sharing of technology and R&D, just as they would a competitor.

The perception of minority-owned joint ventures as an "add-on" in terms of opportunities for exploiting previously accomplished R&D can also be seen in the example of an American firm with a joint venture in Japan. This joint venture has a staff of about forty technical people of its own, plus eight or nine R&D professionals at the Japanese partner's laboratory who are dedicated to the joint venture's research activities. The joint venture originated with the Japanese partner owning a firm in need of technical assistance in a field where the American company had a substantial technical reputation. At the same time, the U.S. firm was looking for overseas investments. Initially, the U.S. partner supplied all of the technology, but today the joint venture does some inhouse R&D, and the Japanese partner's own central research effort is fed into the joint venture. Twice a year, meetings are held with the Japanese to discuss items of mutual interest and to exchange technology (generally via a formal licensing arrangement). Recently, a piece of technology developed in the U.S., but not marketed, was picked up by the Japanese, who are now considering selling it in the United States.

The same American firm owns one-third of a Latin American joint venture, of which another one-third is owned by a major European competitor. Both firms make their existing technology available without modification to the Latin American firm, which selects the technology it will adopt depending upon product line and based upon technological abilities and managerial involvement. This joint venture thus offers new markets for existing technology, rather than an incentive to do new R&D.

Slightly less than one-half of the joint ventures discussed in our interviews appeared to represent full-fledged commitments by the partners involved to participate in R&D support of the joint venture. In every case where the ownership position of these joint ventures could be ascertained, the joint ventures having the transnational corporate partner's R&D support were also those where the transnational corporation held a majority ownership position. An example is a European firm that owns 65 percent of a joint venture in the United States in which R&D was begun to support manufacturing and marketing. As the joint venture grew, so did the technical capacity of the firm, and an engineering development group based on European know-how was added. Since the product line and composition is different in the United States, different capabilities are required, and the American lab is now self-supporting in product development. Process development responsibilities, however, are retained at the European headquarters, and the joint venture laboratory has an R&D agreement with the European partner stipulating that there will be automatic and full licensing of the know-how developed in Europe, and providing for a reverse flow of information from the United States to Europe.

In sum, active R&D commitment by the transnational companies interviewed to a foreign joint venture is predicated on majority ownership by the party concerned. As at least two R&D managers stated, in almost identical terms, "The majority-owned joint ventures get a full flow of information and support from the labs, and all know-how available. The minority-owned affiliates are held closely within contractual relationships. If a foreign partner wishes to do R&D on its own, we will not discourage it—but for us to do R&D for them requires a contract on a specific basis."

Of course, while ownership interest appears critical, the identity of the partner is also important. This is particularly true when the partner is an important competitor in other international markets. In general, the net result of competitors joining together in a joint venture to serve a particular market is to constrain the growth potential of, and information flow to, the joint venture. This is illustrated by two similar joint ventures involving an American firm. In one of these joint ventures, the American transnational holds 40 percent ownership against a local partner's 60 percent, yet supplies the joint venture with *all* of the technology it has in the United States, with only a short time lag for the patent application period. Representatives of the joint venture sit in on technical reviews and actively communicate with the American firm. In the other joint venture, in which the firm holds a 50 percent ownership position and the other 50 percent is held by an important American competitor, the presence of the competitor has severely dampened the flow of information to the joint venture. Although there is some communication between the firm and the joint venture, precautions have to be taken to prevent the joint venture from reporting back to the other partner. Neither of these joint ventures involve R&D by the affiliate.

A joint venture in a developing country between an American and a British competitor with a local third partner provides a second example of the dampening factor of a competitor's presence in such arrangements. This joint venture grew to be a sizable operation, including its own R&D division, but without technical interchange with the partners because of their fear of compromising proprietary information. Recently, however, the American partner bought out the British partner and has instituted a substantial technology sharing effort. According to the manager interviewed, it was the presence of the British competitor that had caused the lack of technological interchange, and now that the competitor-partner was gone the American multinational could fully support the joint venture's R&D activities without risking the loss of proprietary information to a competitor.

JOINT R&D VENTURES

In a number of cases, R&D interaction between firms was accomplished outside of institutional structures. These arrangements took the form of cooperative research activities at one end of the spectrum and a sharing of R&D plans and information at the other. They existed in both bilateral and multilateral forms (with the latter particularly evident among the nonferrous metal firms), and they were entered into to satisfy a variety of objectives.

Typical of the cooperative research ventures existing among firms was the involvement of a European multinational with a smaller European firm in a technical area not within the original R&D programs of either organization. The joint research efforts shared by the two firms are defined by subject area, and all R&D efforts within these areas go into the joint venture from wherever they might arise in either of the two companies, including Central Research. Benchworkers meet at least two or three times a year, more frequently if necessary. A joint science committee guides cooperative research projects, and a project leader provides liaison between the two research teams. Each firm can market any resulting product anywhere, although it is difficult to market the same product under two different trademarks. Market penetration is open without royalty to either firm in any market, including one already occupied by the other.

Another joint R&D venture between two firms was initiated by one of them in order to forestall the entry of a potential competitor. In this case, the firm interviewed had for some time supplied a synthetic raw material to another firm that was considering integrating backwards. In order not to lose a customer, and to discourage the birth of a potential competitor, the firm offered to enter into a joint manufacturing venture for which some initial R&D was necessary. Prior to the establishment of the joint manufacturing venture, however, a joint research project was instituted with a project director in each company and with the temporary assignment of scientists from both companies to the joint R&D project. This arrangement lasted several years, and was quite successful; it is now being terminated and will be picked up by the joint manufacturing venture.

Somewhere between joint R&D ventures and open sharing of R&D programs and plans lies the experience of a pair of electronics firms, one American, one European, who have actually exchanged R&D personnel in order to pursue joint research in a particular product area. This relationship evolved from a twenty-five-year friendship between the firms' respective R&D directors and was carried out without the need for formal coordination.

Several examples were found of research plans and programs shared between firms. One case of bilateral sharing was found in an American engineering company that had a technical exchange with a European machinery manufacturer. In this case, the American firm had made a conscious decision not to duplicate the partner's R&D efforts in specific technical areas. Accordingly, contractual and traditional ties evolved that have led to a rather complete sharing of technical developments. The two firms hold semiannual, formal, research-sharing meetings and enjoy a multitude of intensive, informal meetings, including the frequent on-site presence of a representative of R&D from the American company at the laboratory of the European partner. This results in a loosely coordinated joint R&D program subsuming the individual R&D programs of the two firms.

An example of multilateral research sharing was found among some of the firms interviewed in the nonferrous metals industry. International Research Associations exist in the copper, lead, zinc, selenium, and tellurium industries, and their memberships include private and public sector producers of the particular minerals. These associations support research efforts wherever the best capabilities to tackle the major research questions exist. According to the R&D executives interviewed, competitive advantage in their industry is largely determined by process improvements. Therefore, the R&D activities sponsored by the International Research Associations are in technical areas where the results will not challenge the existing competitive balance, but will benefit everyone. Among such areas are market and new product development, and the control of environmental degradation.

The R&D director of one nonferrous firm stated that any product application R&D performed by his company (which is, to be sure, a very small percent of the firm's R&D activities) is donated to the appropriate International Research Association. The rationale behind this "generosity" is that anything that develops new markets aids the producers, and could not be protected, anyway. His firm gains by being the first on the scene to understand the subject and by developing in-house expertise. In addition, donations of such information tend to increase his firm's international stature in the relevant technological communities. Since the real competitive technical advantages in this industry come in process improvements, membership in the research associations carries very little cost.

Not all of the firms interviewed felt that research sharing between corporations had an appreciable impact on their own R&D programs. One company in particular has had several research-exchange agreements with other firms in which visitations and reports are shared but are generally confined to a specific technical area. How well such arrangements actually work, however, is questionable, because, according to the executive interviewed, "all participants withhold information. As firms expand into each other's markets, they can no longer get along, and the agreements tend to lapse." This firm's participation in such arrangements is a passive one, limited to report-sharing, rather than an active one, with actual research collaboration. A second perspective on passive involvement in research-sharing arrangements was offered by an R&D manager in a chemical firm, who felt that such agreements were "a bonus, although you can't rely on the other companies. They don't cost [because of the firm's passive involvement], and they sometimes yield benefits."

COOPERATIVE R&D PROJECTS

One interview yielded insights into multilateral cooperative R&D, which occurs when a consortium of companies is established for a large single project, such as certain NATO endeavors. In these cases, the firms are not doing research for other companies, but with them. The typical areas of such "systems bidding" arrangements are transportation, water supply, urban reconstruction, and new cities. As civilian and military communications and transportation begin to overlap, even larger systems are being bid for. Also, local subcontracting of system parts is increasingly being required by the purchasing governments, further complicating these arrangements. According to one of the R&D directors interviewed, consortia for systems bidding are likely to increase in number and significance in the future and, hence, future R&D must increasingly take into account the systems requirements of customers, and be prepared to take part in such schemes. In an effort to be prepared, this firm is attempting to enter into cross licensing arrangements with its international competitors, yet it recognizes that communications among R&D labs of independent companies is difficult because of the need to maintain rights to proprietary information and prevent information leaks.

SUMMARY

Joint ventures and research-sharing arrangements are not infrequent, and they are sometimes beneficial to the participating firms. A variety of approaches to collaborative R&D were uncovered during the course of the interviews, but the key determinant of active participation by a firm in such an arrangement is the competitive advantage to be gained by such participation. Foreign joint ventures will be permitted to undertake new R&D efforts when the transnational can maintain control over that R&D—generally, when it enjoys a majority interest in the joint venture, providing both control and profitability. Minority ownership positions and/or the collaborative presence of a competitor apparently pose sufficient risk to proprietary R&D interests to restrain active participation in such situations.

It is interesting to note that the nations more desirous of attracting R&D to local activities—advanced developing countries—are also those insisting on minority ownership positions for the transnational corporations, thus further reducing the chances of TNCs undertaking R&D activities in these countries.

The noninstitutional sharing of research will occur when necessary to enter new markets, or to address the mutual needs of the partners but will not greatly effect the participating firms' individual activities if there is a risk of losing competitive advantage through the sharing process.

Chapter 5

The Diffusion of R&D-Related Capabilities

One reason host governments seek the presence of transnational corporations' R&D activities is to absorb R&D skills into the local economy to stimulate indigenous capabilities by local institutions. The primary, first-order effects of diffusion occur principally within the scientific and technical communities of the host country; secondary effects diffuse among competitors and eventually into all industry. Our study focused on the primary stage, and the diffusion into the host country's industrial and nonindustrial scientific and technical communities.

DIFFUSION INTO INDUSTRIAL SCIENTIFIC AND TECHNICAL COMMUNITIES

Diffusion of R&D orientations, skills, and capabilities occurs in a number of ways—through personnel, publications, and other modes of communication. Major diffusion occurs in hiring and training of professionals for staff positions. Turnover of this professional staff spreads the acquired skills and capabilities within the host country leading to an upgrading of technical sophistication in local sectors. Similarly, the promotion of host-country professionals into R&D managerial positions diffuses skills to the host-country communities, again through transfer of personnel and communication among various positions with consequential exposure to new responsibilities. Training programs at all technical levels within transnational corporations represent yet another means by which host-country capabilities are upgraded. Finally, providing technical assistance to local suppliers (vendors) and customers in order to improve quality, scheduling, and service, to achieve manufacturing compatibility, and to develop markets, opens yet more channels of diffusion to host-country communities.

Employment of Local Professionals

Without exception, all of the firms interviewed recognized the desirability of employing host-country nationals in the professional staff and managerial positions of their foreign R&D activities. In the majority of cases the former had been accomplished. In most of the other instances, substantial progress had been made toward this goal. It was not unusual for a firm to report that 100 percent of its professional R&D staff abroad were host-country nationals.

In several cases, firms noted that they typically began foreign R&D operations with their own nationals and then switched to employing host-country nationals. As an example, one American firm noted that its electronics lab in Britain "at first had all U.S. personnel, but most had left by two years, and all had left by six years. The manager of that laboratory for its first nine years was an American, which is generally the case for an initial overseas laboratory manager. Very few Americans, however, are now at our overseas operations."

Acquiring suitable local R&D staffers, however, no matter how desirable a policy, may not be easily accomplished. According to one European firm, the Latin American host country to one of its overseas R&D facilities is "really impossible as a country for central research activity. We have had 100 applications for positions in research-type activities, but these have shown themselves to be without adequate training, and without a professional attitude or curiosity. There appears to be no adequate educational facility in this country to service the types of people they need. We would consider a design department, but there are not yet adequate students coming out of the academic community to handle even that." Several American firms with numerous R&D activities overseas have reported similar experiences, and have reduced, or in some cases reversed, efforts to set up activities in these countries.

Anticipated diffusion from turnover was not in evidence, since the reported turnover among host-country R&D professionals was quite small. In those few cases where some turnover was mentioned, it appeared that the departing R&D staffers moved to a competitor, typically another transnational corporation, rather than establishing their own business, joining a university, or taking a position in a government lab. This is not the pattern in some developing countries (LDCs) where government jobs are preferred, and professionals remain in industry only when no civil service position is available.

Employment of Nationals in
R&D Management

Most firms employ host-country nationals as R&D professionals, but positions in management of foreign affiliate laboratories of transnational companies (TNCs) are more often filled by expatriates. According to one U.S. R&D executive, "Language is an extremely important consideration in selecting an R,D,&E director. Nationals in English-speaking countries will become directors more quickly than will those in non-English speaking nations, because headquarters will be more comfortable relating to someone who shares their culture." Although no other executive stated this opinion so emphatically, examination of the actual staffing practices tended to support such a view. In a number of instances, a local national who had worked in the firm's headquarters for some time was chosen to return to his homeland and become the laboratory director. Such a scheme takes advantage of established communication and trust relationships, yet avoids the appearance of foreign domination. This latter course does not, however, always work. One U.S. firm, faced with a need to establish some control over an European acquisition that was still dominated by its entrepreneur-founder, sent a host-country national who had been employed for some time at the firm's headquarters laboratory back to his home country to run the lab. "It was a disaster! He built a fence around the laboratory, fought continuously with Manufacturing and Quality Control, and eventually had to be fired." The next lab director was chosen directly from the staff of the acquired group and has worked out well.

The opinion was also expressed that it was sometimes necessary to employ home-country nationals as directors of overseas laboratories in order to ensure an appreciation of the role of R&D in the foreign affiliate. One particular firm had a laboratory in a developing country affiliate that was dominated by sales personnel with little understanding of the need for research. Accordingly, the firm believed that it was "necessary to have a European as the top man in the firm in order to develop an R&D orientation. However, the European selected was also sales-oriented, and so yet another European had to be put in as R&D director of the foreign affiliate in order to match the strength of the first European in top management." Clearly, nationality is not by itself sufficient for R&D success in foreign affiliates.

A handful of the firms interviewed were actively trying to move towards an international R&D staff. One manager stated that "while almost all of our overseas researchers are local nationals, the directors of our overseas R&D groups are not necessarily local nationals. The nationality of a laboratory director really does not matter to us if he or she is the right person. We don't have a corporate policy on this matter, and only if all other things were equal would we consider local nationality as an important determinant of preference for a laboratory director." Another firm expressed the hope that "eventually we'll reach the point where we will fill the technical directors positions with the best person available, regardless of nationality." These two particular firms had already made considerable progress in promoting foreign nationals with technical backgrounds through the hierarchy of corporate administration in the United States.

Similarly, a European transnational has adopted as a personnel policy "to become as multinational as possible." At present, workers from forty different nations work at company headquarters. Furthermore, this firm shifts personnel in and out of its home-country office in order to internationalize workers and their families, the objective being the development of "both individual and family loyalties to the company and to each other." One result of this policy is enhanced mobility of personnel among the various development groups and between R&D and production. The direction of the process is largely one-way, however, from R&D to production, with R&D becoming a training ground for university-trained new personnel who are not yet "industrially oriented." Rarely are personnel in other functional areas of a company sufficiently prepared technically (or in career orientation) to move into R&D work or management.

R&D as a Training Ground

The use of R&D as a training ground for personnel to be moved to other areas of the company is not unique. All firms reported extremely low turnover of R&D personnel, both at home and abroad, but R&D staffers are frequently sent to other activities within the company. A Japanese firm, for example, regards its R&D laboratories as "sources of additional engineering capabilities when the business needs more on the plant floor." In 1978 it transferred 20 of its 300 R&D personnel to other company operations. An American transnational corporation in the metals industry was experiencing a rather high (15 percent) annual technical staff turnover in its Latin American laboratory, of which approximately one-half was due to promotion of technical people into local operating facility positions, "a fact [that the executive interviewed observed] reflected the role technical work often plays as training for administrative positions." According to the director of a U.S. chemical-based corporation: "We're never satisfied with the number of technical people moved into nontechnical positions. However, when a technical staffer is making more than $30,000 U.S. per year, or is more than forty years old, it becomes extremely difficult to move him or her to a corresponding nontechnical position." The personnel ties between R&D and the operating facilities that can be gained by personnel transfer have led one European transnational to begin shifting each year five percent of its R&D staff into divisional testing facilities and manufacturing. This keeps the R&D group's average age at thirty-five, which reflects the company's belief that "younger people are more creative in research, and older people need to be placed in applications."

While personnel transfer from R&D to other corporate activities appeared to be common among the firms interviewed, the opposite was true with the movement of R&D personnel among the R&D laboratories of the same corporation. A variety of reasons were given for this lack of international technical mobility, but basically these reasons all addressed the cultural differences among the firms' affiliates. Professionals are themselves unwilling to make a cross-cultural move, and in some countries this reluctance means that a professional cannot be transferred from one city in the country to another (Rome to Milan, or Rio to São Paulo, for example). Also frequently mentioned were salary differentials between countries and visa problems, which made international transfer awkward. Thus, while everyone appeared to agree with the adage that "the only effective way to transfer technology is to transfer the people who did the R&D," most firms tended to limit such personnel transfer to only short durations at the staff level, with third-country transfer occurring, if at all, when the individual reaches the general manager level.

Short-term transfers on a limited scale do, however, appear to be a primary means of transferring research skills. Typically, such activities are limited to one or two individuals from each foreign laboratory, spending six to twelve months per year at the headquarters laboratory. While the experiences and motivations vary greatly from firm to firm, in most cases two primary objectives underlie such short-term transfers: the opportunity to teach workers research skills and the desire to establish personal contacts that will facilitate future information transfer.

The use of short-term transfers, or visitations, also represented an important means of achieving organized (though unstructured) training. With only a few exceptions, the firms reported that any formal training programs for R&D personnel were the responsibility of the overseas laboratory and most likely did not exist as distinct from company-wide programs. Given the relatively high educational levels of R&D staffers, this lack of commitment to corporation-instituted formal training programs is not surprising. Attempts were being made, however, particularly in those activities located in developing countries, to inculcate sound research practices in the local staff. In one firm, a two-person R&D team visits a foreign laboratory in a developing country for a period of one month every quarter. The work they perform there is typically unsophisticated product blending experiments because they are constrained by the lack of sophisticated instrumentation available locally. The purpose of these trips, however, is not to make technological advances but to perform an educational function: "to show them how to do this and that."

One of the benefits of visitations as a training vehicle is that they create a cadre of interchangeable R&D staff people, so that internal personnel transfers in times of emergencies are possible. In addition, such training is necessary if international career paths are ever to be fashioned. During the course of an interview at one firm, a phone call received by the technical director revealed his detailed knowledge of personnel capabilities at a number of foreign R&D facilities and a willingness to transfer people internationally to meet the firm's needs. Equally impressive was evidence of the director's sensitivity to the trauma of international transfer and empathy for the individuals involved and their families.

The desire to develop personal contacts with overseas staffs in order to facilitate communications was evident in most of the companies interviewed. Most often this occurs at the managerial levels, but it is frequently a factor in staff transfers. Though not all R&D personnel, one Japanese firm reported that over 200 of its people leave Japan each month for overseas trips of a week or more in duration to visit field offices, and take training courses. This particular firm, however, derives 95 percent of its income from overseas operations and thus many of its overseas trips are for familiarization with foreign cultures and information sources rather than for technical training. Another transnational corporation recently relocated forty Japanese families to a U.S. facility for a single case of specialized product development. When they returned to Japan, the team had the entire project well in hand, something the transnational firm does not believe the team could have accomplished had it attempted work in Japan. During the period of transfer, some sixty other Japanese were also brought to the United States for technical service training. Far more common is the experience of an European firm where a formal overseas laboratory sends one executive and his family to the European headquarters for a year, each year. An American firm with a world-market orientation has successfully used very short duration visitations, where the traveling executives actually stay with the families of the staff being visited, to establish the personal rapport necessary to facilitate communications.

Diffusion to Suppliers and Customers

One of the most direct ways for transnational corporations to diffuse R&D skills and capabilities into host countries is to upgrade the capabilities of local suppliers and clients. Quite a number of the firms interviewed had tried one form or another of this method, more often than not within developing countries.

Direct technical assistance to customers is to be expected in any type of high-technology commercial transaction. In the interviews we conducted for this study, we were primarily concerned with the role R&D played in such assistance schemes when the clients were located in foreign countries. One American chemical firm, for example, boasted of actually establishing the plastics industry in a large and advanced developing country. It had done this by sending "applications teams" into the field to demonstrate what could be done with plastics, and how to fabricate products out of plastics. While the ultimate objective of this effort was to create a market for the firm's expoxy resins, it had the effect of significantly changing local products and production techniques. A European transnational hosts a technical school with programs lasting from one to six months for personnel with different backgrounds from all over the world. They learn the systems that are being purchased from the company, including specifications of products. The benefit to the firm of this training is in the contacts made with the customers. Less spectacular, but perhaps more representative, are the technical seminars presented at foreign customers' facilities by an American transnational corporation's R&D representatives who, although based in the United States, make regular circuit tours through Europe in support of Marketing.

Technical assistance was also provided by the transnational corporations to their foreign suppliers. Most of this assistance involved establishing improved quality control practices and locating sources of raw materials required by the vendors. In a number of firms where agricultural products were important as raw materials, elements of the transnational corporations' R&D groups were active in working with the host-country agricultural community to develop improved farming practices and crops. Due to the small sample of such firms, few conclusions can be reached, but some indications point to the conclusion that the more decentralized the influence patterns within the firm, the greater the willingness to take raw materials as given and adapt the production process to them, rather than attempting to upgrade the raw materials.

Some firms go quite far in providing technical assistance to their foreign suppliers, as shown by a transnational chemical corporation that had a local supplier of a fairly hazardous material in a Latin American country. The route between the supplier's facility and the transnational's plant was a dangerous one, part of which included a steep section of road leading into a village, representing an appreciable potential for disaster. To reduce the likelihood of mishap, the transnational built specially designed trucks resembling tanks, and despite the considerable expense involved, these trucks were given to the supplier. This, of course, is unusual, but it illustrates the willingness of some of the transnational corporations we interviewed to employ their R&D groups to provide the assistance necessary to successfully integrate local suppliers into the TNC's production and distribution process.

DIFFUSION TO NONINDUSTRIAL SCIENTIFIC AND TECHNICAL COMMUNITIES

The R&D activities of a transnational corporation are also a source of diffusion of scientific and technical capabilities to the nonindustrial scientific and technical sectors within a host country—particularly universities and government institutions—helping to form a community-of-interest.

Diffusion to Universities

The importance of host-country universities to transnational corporations has already been demonstrated by the corporations' interest in local university strength when considering potential locations for foreign R&D activities. For some firms the presence or absence of a strong university near a candidate foreign location was *the* significant influencing factor in determining the foreign location. Interestingly enough, this perspective was not necessarily limited to high-technology firms, but was important to a wide range of companies. Furthermore, our interviews indicated that some cooperation existed between transnational corporate R&D activities and foreign universities. For the most part this took the form of individual consultative arrangements with faculty members, or "moonlighting" by corporate R&D staffers as adjunct professors. These activities were quite common, particularly the latter, but they occur largely by chance, and are more predominant in the developed countries. In the pharmaceutical industry, however, local medical requirements often result in heavy reliance upon local physicians and university faculty for the performance of clinical trials.

Several firms had made organized efforts to develop relationships with foreign universities, particularly in Europe and Japan, where universities were seen to be a relatively more important source of ideas for R&D than were their counterparts in the United States. According to one American pharmaceutical firm, the importance of universities to its overseas R&D laboratories might be a result of the typically smaller size of foreign R&D units and, consequently, the lack of both depth and breadth in the laboratory's inhouse expertise. Several companies believed that foreign universities were often the best and/or most accessible source of information regarding local scientific and technical activities and, hence, close cooperation with these universities was warranted as a scientific and technical monitoring effort. One such approach led the local affiliate of a transnational corporation to host a two-day "Professors Day" each year, where twenty-five local professors are invited to the laboratory for a seminar during which they have the opportunity to explain their current research. This same firm has purchased and donated laboratory equipment for foreign university use, brought Latin American faculty to the United States for training in toxicology research, and sponsored university co-op programs, research opportunities, and grants. A second American transnational has sponsored joint university-corporate appointments in Mexico and Great Britain, and has offered fellowships, visiting professorships, and postdoctoral positions to foreign nationals, and sabbaticals to its own staff. One European transnational had even established an R&D laboratory in another European country in order to share research with two local universities under a basic research cooperation agreement.

Several of the transnational firms stated that they had subcontracted or sponsored R&D at foreign universities, even though they were not physically located in that country. Similarly, an American firm maintains worldwide university contacts in its peculiar area of research, which tends to take on the appearance of a "closed club" because of the small number of people working in this area.

Diffusion to Government Institutions

The relationship between host governments and transnational corporations regarding the performance of R&D is considerably more complex than the relationships between these firms and foreign universities; this chapter will examine only the scientific and technical interchange between the transnational corporations and their hosts.

Although only a few firms emphasized their scientific and technical cooperative activities with host-country governments, the impression conveyed was that such activities, while not unusual, were not planned, either. By and large, the activities in which these companies assisted host-governments were those where they could contribute their special expertise, and where such efforts would present, or reinforce, a positive image of the firm's presence. For example, a European host-market firm in the food industry helped establish, via subcontract, quality control capacities within several government organizations in Latin America. Similarly, an American food firm donated some of its overseas R&D staff's services to local government advisory committees concerned with that sector. A host-market chemical company sent some of its scientists to a Latin American country for several weeks to assist its Environmental Protection Agency, and a pharmaceutical firm reported that three Latin American countries have adopted its corporate quality assurance standards as their own government regulations.

More ambitious are the efforts of a number of firms who have actively collaborated with foreign governments to establish local scientific and technical capabilities. These activities are fewer in number than the more passive interactions described above; in these few cases, some overt host-government pressure appears to be present. For example, one host-market firm, dependent upon agricultural raw materials for its production input, maintains an agronomy group in R&D which, in one case, is providing technical and manufacturing expertise to a Middle Eastern government competitor, which also controls imports. Since the transnational corporation enjoys a prominent position in what is considered an important market, it behooves it to provide its government competitor with improved technology and production practices in order to maintain the friendship. Similarly, a European transnational has built a school for technicians in two developing countries in cooperation with the local governments and has turned them over to the governments.

Only one firm told of performing R&D for a local foreign government—work for the Canadian Research Council on a product the Canadians wanted to develop that the company felt would indirectly benefit it through reduced intermediate product prices. The firm's Canadian subsidiary carried out the research phase, with the project being moved to the United States for pilot plant operations. Eventually, the firm decided that the project did not promise enough return for further involvement; it terminated the work, relinquishing all of the information developed to the Canadian government.

SUMMARY

Evidence does exist of the diffusion of R&D capabilities and skills from transnational corporations pursuing R&D abroad to host-country industrial scientific and technical communities. The magnitude of such diffusion is not very large, however, nor is it typically the result of conscious decision. Technical assistance has been provided to local suppliers and customers and occasionally even to competitors when it was necessary for smoothly integrated operations. Upgrading of the scientific and technical skills of the indigenous population typically did occur, but only among the already highly educated—to be expected, since only the most highly educated are employable as R&D staffers.

Turnover among indigenous R&D staffers was minimal, and did not appear to promote skill diffusion to the host-country scientific and technical communities as much as it led to the migration of skills and capabilities among competing transnational corporations. R&D positions were clearly, however, a promising pathway to general administrative positions in the managerial hierarchy of the subsidiary, and occasionally within the transnational corporation itself.

Although numerous examples of the diffusion of scientific and technical capabilities from transnational corporations to the host-country's nonindustrial scientific and technological communities were reported there was no evidence that such activities were frequent or significant; the impression given was one of chance and confluence of interests, rather than the result of planning. While host-country governments have seen indigenous scientific and technological institutions and capabilities bolstered through interaction with multinational corporations, this has not occurred as a result of considered government or corporate policy.

Chapter 6

Impacts of Governmental Policies on R&D Location and Performance

The impact of host-country governments' policies on transnational corporations is of critical importance to the whole issue of the corporations' pursuit of R&D abroad. Despite the potential significance of governmental policies, R&D managers reported that, in the large majority of cases, foreign governments had had no effect on their foreign R&D decisions. Yet a closer examination of the situation showed that this answer is only partly correct. When the rewards to be gained by pursuing R&D abroad outweighed the costs incurred by host-government interference, the firm typically put up with significant host-government impositions. Conversely, when the rewards of doing R&D in a particular market were not perceived as being worthwhile, no amount of host-government enticement was sufficient to influence the R&D decision. Thus, host-government actions in developed and advanced-developing nations were far more likely to have a significant impact on a firm's conduct of overseas R&D than were the actions of less developed countries' governments. Complicating this asssessment, however, is the fact that the "rewards" are sometimes the removal of onerous regulations or the opening of government purchases to the TNC, rather than the benefits provided to R&D.

Host-country governments influence a transnational corporation's pursuit of R&D by use of positive or negative inducements. Among the former are tax advantages and monopoly privileges, subsidies, government purchases, patent privileges, etc., while the latter include import restrictions, local registry conditions, price controls, and joint-ownership regulations.

POSITIVE INDUCEMENTS

Tax subsidies appear to be the most popular positive inducements offered by host-country governments in their efforts to attract the R&D activities of transnational corporations. They are also undoubtedly the most desirable, according to the R&D managers interviewed in our study. Yet even so, we failed to identify even one R&D location decision where the availability of tax incentives was more than simply "icing on the cake," once the location decision had been made on the basis of other, more important, considerations. As an example, the tax incentives offered R&D performers by the Canadian government had attracted the serious attention of many of the managers interviewed. But in every case where a decision was actually made, the firms decided against a Canadian location because the Canadian market was so similar to the U.S. market that the economies of scale from a single centralized lab outweighed the Canadian incentives. In those cases where a Canadian laboratory was established, it was to take advantage of unique Canadian market factors, or as a result of acquisition, rather than because of the incentives. As evidence of this, all the Canadian laboratories identified in our study had R&D programs oriented to distinctly Canadian problems (such as extreme cold) or tastes, and typically had little overlap with R&D programs in the parent country.

The general ineffectiveness of subsidies can be seen in the case of a transnational chemical corporation that refused every offer of subsidies when locating an R,D,&E facility in a certain Latin American country, even though two other transnational competitors had demanded a tax subsidy tied directly to every dollar of capital investment for R&D brought into the country. This firm believed that the market was attractive enough to warrant the establishment of local R&D activities, provided it could enjoy the right to bring lab equipment into the country duty-free. Agreement was reached on this issue, and the company is presently involved in establishing a large R&D facility there.

The protection of the transnational corporation's investment in intellectual property through respect for patent conventions, although a positive inducement, appears to be more significant as an indication of host-government sympathies towards business than as a real consideration in the foreign R&D location decision. With the exception of the pharmaceutical firms, most of the companies interviewed appeared to consider patents primarily of importance in "staking claims," advertising for leverage in joint ventures, and, as noted by one executive, to "guarantee that at least no one can prevent us from using our own processes." This individual stated that "when you start moving abroad and want to influence customers, the number of patents you hold is often regarded as an indicator of corporate technological capabilities." Among pharmaceutical firms, patents are quite important to the protection of the R&D investment, and several companies indicated that Italy's former pharmaceutical patent prohibitions had excluded that country from further consideration as a location for R&D activities.

In general, positive inducements offered by host-country governments to attract R&D were rather limited both in incidence and in effectiveness. Canada, Italy, and Brazil were mentioned frequently as offering such inducements, but not as locations made attractive by these offers. One executive stated quite emphatically that the only positive inducements his firm would recognize were those policies that "make for a stable, hospitable investment climate, and which provide assurances of a community of interests among all parties involved." This does not, generally, mean the offering of tax subsidies and the like, for these will not work in situations where, on the basis of market conditions alone, the firm would not do R&D in the first place, and they are unnecessary where the market conditions are already such to attract R&D. Perhaps the most positive incentive a nation can offer, therefore, is a clear and mutually advantageous set of rules for doing business. Some of the firms interviewed suggested the Soviet Union as a model for such behavior.

NEGATIVE INDUCEMENTS

Negative inducements appear to be far more effective than positive inducements in influencing the R&D decisions of transnational corporations. The key to their effectiveness, however, is the attractiveness of the particular market, and whether or not the firm is presently interested or engaged in doing R&D in that market. Consequently, developed nations, with attractive market opportunities, appear to be successful in employing negative inducements to compel the location of R&D activities in their country. Developing nations, with less attractive market opportunities, can employ negative inducements successfully only when the transnational corporation has already determined that it could be profitable to perform R&D. Governments can alter expectations of profitability by use of price controls, exchange controls, import restrictions, export requirements, protection of patents, etc. By forcing reassessment of *potential* profitability, the government can "seduce" or "blackmail" a company into establishing an R&D activity, but only if returns remain attractive.

Among the developed countries, Japan and France are most frequently mentioned as successfully compelling the establishment of local R&D activities through the exertion of substantial pressure on the firms. The establishment of local R&D activities by a TNC is often set as a precondition to gaining market access in these countries. In some cases, the results can be quite significant. One firm established a full-spectrum research center in France with a staff of 100 scientists in return for permission from the French government to acquire a French firm. According to several executives interviewed, however, this facility is viewed as representing "nothing more than a drain on corporate profits for the next twenty years with little practical good coming out of it." A similar assessment was made of a much smaller facility that was located in Japan as a direct result of Japanese government pressure. In this case, Japanese government regulations required in-country performance testing of the petroleum products the firm was selling. An R&D laboratory was established to satisfy the regulations, but the "laboratory" is specifically limited to the task at hand, and has no skills higher than engineering.

An advantage possessed by developed nations arises from their scientific and technical capabilities; consequently, a firm pressured to locate there may actually profit from that location. As an example, one firm that desired to sell in the French market was persuaded by the French government to establish a laboratory even though it had no French manufacturing operations. Despite this handicap, the R&D manager reported that "the quality of French scientific and technical resources available, and the importance of the Common Market, will undoubtedly lead to an evolution of the French laboratory into doing some product discovery research." Similarly, another firm reported that Japanese requirements that the firm perform local research on carcinogens and mutagens led to the establishment of an unwanted R&D presence there. In order to achieve critical mass, the laboratory has been expanded in size and responsibility, and it will eventually be used to reduce duplications of R&D efforts company-wide. Once again, this is possible only because of the sophistication of the Japanese scientific and technical communities.

The imposition of restrictions on the ability of the transnational firms to repatriate their profits has been frequently employed by both developed and developing countries to induce the establishment of local R&D activities. The intent of such restrictions is to force reinvestment of capital in the country. Presumably, one of the investments to be considered is R&D, and we found a few instances of the desired results. One European transnational reported that as a result of the Brazilian government's prohibition on the payment of royalties on technology, it had decided to begin local R&D on a low-technology product that was really of no particular importance to the company itself, although it was being sold by its Brazilian subsidiary. The firm is not sure that it will get anything at all from the investment.

A similar situation occurred in Spain, where a firm which, when faced with the Spanish government's prohibition on royalties from technology licensing, established a Spanish R&D lab. This was done despite the Spanish subsidiary's staff's lack of confidence in the action. The result in this case has been a positive one, however. The Spanish R&D lab specializes entirely in one product line, and has gradually been upgraded until there is presently very little difference between the quality of research being performed at the Spanish laboratory and that being performed at headquarters. In fact, there has even been some reverse technology transfer from Spain to the corporate headquarters. In at least one company interviewed, however, funds that cannot be repatriated are used to host the firm's yearly international research meeting; thus the firm can claim that this money is being expended locally on R&D, although no research actually takes place.

In another case, a firm faced with paying higher taxes to assist the development of indigenous R&D laboratories was given the option to spend the funds instead for its own local R&D. In addition, the host government required semiannual reports on these R&D programs, separation of the local laboratory from mere support of manufacturing operations, and an understanding that the firm's local laboratory would be pursuing work more sophisticated than quality control. Given that this particular market, in an advanced developing nation, was a desirable one, the firm decided that establishing its own R&D laboratory made far more sense than supporting a government R&D institution through the payment of higher taxes. Accordingly, it has established a twenty-person laboratory, three-quarters of whom are professionals (BA or higher). At present, the laboratory is pursuing a project involving local materials for raw material processing and a project adapting styrofoam to local conditions.

It is among the developing countries that we found the most willingness by host-country governments to use "border-closing" policies such as local content requirements and import substitution schemes to induce transnationals to locate R&D in their countries. One example of how such schemes work was provided by a transnational electronics firm with a subsidiary in Latin America. This company originally established its Latin American presence by importing its products for sale in the local market. The host government stopped this in an effort to encourage local manufacturing; in response, the firm began to import component parts and assemble them locally. This, too, was eventually prohibited by the local government's import substitution policies, and the firm was forced to go to a local manufacturer and redesign the product with new specifications. Consequently, the company established a local R&D group that is now growing fairly rapidly.

A similar result arose out of border-closing actions in 1972 by the government of an advanced developing country that for some time had been pressing a firm to do R&D locally. Eventually, the government promulgated import restrictions on certain materials vital to the company, including an absolute prohibition on the importation of the one ton of benzene that represented the firm's total annual consumption of that chemical! As a result, the firm closed down some of its European research activities on organic chemicals and moved the operations to the developing country. Recently, apparently in response to the firm's willingness to pursue R&D locally, the host government has relaxed its import prohibitions and allowed the firm to acquire necessary materials overseas. The local laboratory, however, is continuing to investigate local medicinal sources— looking for new plants, drying them, and isolating their principle active elements. In addition, the laboratory now has a complete pharmacology unit of 110 persons (22 Ph.D.'s), performs antibiotic screening, has a synthetics program, and does joint development work with the parent company on derivatives.

As a further example of the potential effectiveness of border closings, a consumer products firm reported that, because of import restrictions in an advanced developing country, its R&D group had worked with farmers to develop an acceptable local product for raw material in the firm's production process. Over a period of 20 years, the overseas laboratory had evolved to the point where it is now doing substantial applied R&D.

While the results of border-closing policies by host governments discussed above are not unusual, more typical is the case where the corporate response to import-substitution and local-content requirement policies is the temporary visitation of headquarters R&D staff personnel, who stay only long enough to effect whatever modifications are called for, and then depart. Typically, the local scientific and technical infrastructure is generally regarded as not sophisticated enough to undertake the necessary adaptation, and no thought is given to permanently improving the local R&D staff because, once the adaptation is accomplished, there is no continuing need for those services. Hence, border-closing policies tend to result in one-shot technical fixes administered by headquarters personnel, with the only lasting result to the host country being a less desirable product from the multinational corporation's perspective than that originally produced. Furthermore, the border-closing actions will often gain the host country the reputation of being a "difficult place to operate," thus conceivably discouraging investment by other potential entrants.

A further result of border-closing policies is the creation of local monopolies for producers protected by the trade barriers. At least two examples were found of transnational corporations who under normal competitive conditions would have been performing R&D in the local market, but were doing none, since they held a local monopoly position as a result of border-closing policies. As one executive put it, "We're against border-closing policies when we're not producing in a market, but we're all for them when we're already established there."

It seems clear that the employment of negative inducements such as border-closing policies and profit-repatriation restrictions, although deplored by the transnational corporations, are effective when the country imposing them represents an important market, and the transnational firm has a host-market orientation. In general, the cases related indicate that the most successful impositions of negative inducements occur in developed and advanced developing nations, and, particularly in the latter instance, when the firm has already established some form of in-country scientific or technical activity.

SUMMARY

Although the standard response to our question, "Are you receiving pressure from foreign governments to locate R&D abroad?" was "No," the facts appear to be different. During our interviews fourteen of the American firms and five of the European firms told of some foreign government action that might reasonably be construed as being a form of pressure. Furthermore, as Table 6.1 shows, nineteen laboratories were identified as having their origins in host-government pressure; the countries where such pressure was most evident were Brazil, France, India, and Japan.

The answer to our second question, "Are foreign government inducements/pressures successful in attracting transnational corporate R&D activities?" was most often also "No," but again the facts suggest otherwise. From the evidence presented in this chapter, one can argue that while host-government interference is not as important as market potential in determining foreign R&D locations, it sometimes is more important than the availability of local scientific or technical personnel. By and large, negative government inducements appear to be effective in stimulating some R&D presence in developed and advanced developing countries. However, the value of such activities has yet to be proven; a few have grown into useful laboratories, but the majority have not as yet, and some have proven useless—a cost to both company and country.

Table 6.1. Specific Instances of Host Government Inducements on R&D Location by Transnational Corporations

Inducement	*Country*	*New Lab Established?*
Restrictions on:		
Profit repatriation	Brazil	Yes
	Colombia	No
	Egypt	Yes
	France	Yes[a]
	India	Yes
	India	Yes
	Spain	Yes
Import restrictions	Brazil	Yes
	Brazil	Yes
	Brazil	Joint venture
	Brazil	R&D but no lab
	Brazil	No
	Ecuador	R&D but no lab
	France	Yes[a]
	India	Yes
	India	No
	Mexico	Yes
	Mexico	No
Clinical regulations	Italy	Yes[a]
	Japan	Yes
	Japan	Joint venture
	Philippines	No
Government control		
of the market	Iran	R&D but no lab
	Soviet Union	R&D but no lab
Acquisition permission required	France	Yes
	France	Yes
Drug price control	France	[b]
	United Kingdom	Had some facilities
Tax subsidy	Canada	Yes
	Canada	Expanded lab
	Canada	No
	Canada	No
	Canada	No
	Canada	No
	Canada	No

Table 6.1 *(continued)*

Inducement	Country	New Lab Established?
	Canada	No
	Italy	Yes[a]
Taxes to support government R&D, or forced investment in local R&D labs	India	Yes
Joint venture alternative	India	No
General persuasion	Canada (Quebec)	Eventually
	Brazil	No
	India	No

Source: Interviews.

[a]Combined inducements.

[b]In this case, the firm had not made a decision when interviewed and could not indicate the probable effect.

Chapter 7

Summary and Conclusions

The purpose for undertaking this study was to learn more about how the transnational corporations approach the issue of performing R&D abroad; how they decide to establish foreign R&D activities; how they select particular locations for their foreign R&D activities; how these foreign R&D activities are coordinated and controlled; and what efforts these activities have on the host country in which they are located. Our survey of the literature and the numerous discussions we had with TNC executives in the preliminary stages of this study indicated that only recently has the topic of the transnational corporations' performance of R&D abroad received much attention. Lately, however, interest in this subject has grown considerably as a variety of different policymakers recognize the potentials and problems posed by the next logical step in technology transfer: the transfer of R&D.

The interviews conducted for this study clearly indicated that there is considerable R&D activity by transnational corporations in locations other than their parent countries. Furthermore, this activity is located in a reasonably large number of countries (thirty) and includes many instances where new product research responsibilities have been assigned to R&D groups operating abroad. Thus, there is good reason for continued and heightened interest on the part of policymakers in transnational corporations as they face the need to fashion their own R&D strategies, and on the part of potential host-country governments as they attempt to attract the R&D activities of transnational corporations and, once there, to encourage them to pursue R&D activities of value to the host country. Parent countries of the transnational corporations will remain concerned over the loss of international technical competitiveness, which must be balanced against desires to transfer technology to developing countries. The conclusions presented in this chapter address the concerns of these various policymakers.

CONCLUSIONS

The Pursuit of R&D Abroad by Transnational Corporations

Recent statistical evidence has shown that about 15 percent of U.S. corporate R&D spending is devoted to the pursuit of R&D abroad, and that the number of firms accounting for this spending is quite small and limited to the nation's largest corporations.[1] This finding is consistent with other studies, and with the conventional wisdom acquired in conversations with observers of industrial R&D activities; however, there is some danger of misinterpreting it if one relies solely upon the statistics. While our interviews revealed nothing that would challenge the National Science Foundation's estimates, we did find that many, if not most, of the companies in our sample (firms selected because of their foreign R&D activities) were more than just marginally involved with R&D abroad.

For the most part, the foreign R&D activities of the firms interviewed in this study were smaller in size and more restricted in scope than the R&D activities these firms pursued at home. Furthermore, the decision to initiate R&D activities abroad is an important decision that is typically not made in the normal course of doing business but, rather, represents a departure in the firm's operations. There are, of course, some exceptions to these observations. The foreign R&D activities of the "world-market" firms are often of a similar size and mission as their domestic counterparts and, in fact, are established with the full intention of commanding worldwide responsibilities for certain areas within the firm. Typically, however, our interviews suggested that most firms would prefer to do their R&D in one centralized location, if possible.

That many firms cannot pursue their R&D in one centralized

location often leads to their establishing these activities abroad. The reasons for this are varied, but appear to be strongly related to the firm's market-orientation. The firms most adamant about centralized R&D are those primarily interested in serving only their domestic market. To the extent, however, that these firms needed to have operations abroad for extracting raw materials or performing assembly operations, they also came to need technical service functions abroad. Often, these technical assistance activities evolved into R&D groups with some process-related responsibilities.

Firms involved in the international marketing of goods and services designed to satisfy local styles and tastes have an immediate reason to do R&D abroad: they need to be as close to their markets as possible. These "host-market" firms, unlike the firms with "home-market" orientations, are willing to grant their foreign affiliates considerable responsibility for new product research.

Firms with world-market orientations become involved with R&D abroad in an effort to acquire the best international scientific and technical talent. Unlike the firms in the other two market-orientation groups, the world-market firms typically establish their R&D abroad without regard for the location of their existing international operations. They are much more attracted to concentrations of knowledge and talent than to market volume. Once established, the foreign R&D operations of these companies often become world leaders in certain technical fields.

There are several implications here for policymakers interested in the foreign R&D activities of transnational corporations. First, of course, is the recognition that these firms are doing R&D abroad. Second, their level of foreign activity is closely related to their market philosophy. Industries that manufacture and sell consumer products that are strongly influenced by local styles and tastes, or products whose characteristics are determined by locally prevalent natural or physical geographic or geophysical conditions, are much more likely to pursue R&D abroad than are other industries. The primary inducements for these firms to locate R&D in a particular market are the economic attractiveness of that market and the degree by which the market differs from other markets where the firm is already doing R&D. The resulting R&D activities in such locations will be small, usually limited to applied R&D, with some new product development for the local market, and will often be somewhat isolated from the other R&D activities of the parent firm.

A third important conclusion drawn from the interviews concerns the transnational corporations' willingness to support R&D in joint manufacturing ventures. The firms interviewed expressed their willingness both to enter joint manufacturing ventures and to support these ventures' R&D. This willingness is strongly related, however, to the ownership role accorded the transnational partner in such arrangements. The firms in our sample clearly lost interest in active R&D participation in joint-venture activities when they did not possess a controlling ownership position in the joint venture.

Management Styles and Foreign R&D Activities

The relationship that exists between managerial control and the pursuit of R&D can be seen by comparing the managerial styles of the firms in the sample with their foreign R&D behavior. There was basically an even division among the firms interviewed between those favoring a centralized from of control over R&D and those favoring a decentralized management style. The former tend to have rather elaborate reporting, budgeting, and communication systems, and exhibit a willingness and capability to allocate the corporation's R&D resources on a worldwide basis. The decentralized firms, on the other hand, are considerably more informal in their coordination and control behavior, and view the R&D activities of their foreign affiliates with almost an air of detachment.

Although most of the firms interviewed had central corporate R&D laboratories, the role they played in fostering the firm's technical growth varied as a function of management style. Among the centralized firms, the central corporate laboratory played a major role in the pursuit of R&D programs and projects important to the firm's mission. The ranking R&D executive was often also the laboratory director and was intimately involved, often down to the project level, with the worldwide R&D activities of the firm. His or her counterpart in the decentralized firms was much more of a staff advisor often not directly involved with actual R&D projects, but a reviewer of the R&D capabilities of the firm's semiautonomous business units. The central R&D facilities of the decentralized firms typically address projects of a long-range or new business nature rather than supplementing the R&D efforts already existing within the firm.

The most significant consequences of managerial style appear to occur with respect to patterns of invention and innovation. The complexity of the management control systems employed by the centralized firms result in time delays as research proposals pass through the levels of the corporate hierarchy. These same comprehensive reviews, however, allow for the development of a corporate technical plan and the worldwide allocation of the firm's scientific and technical resources to support that plan. Of particular interest to foreign policymakers is the apparent greater willingness of centralized firms to establish foreign R&D groups by direct placement rather than the more laborious process of allowing them to evolve from technical service activities.

Decentralized firms, while bestowing considerably more freedom on their affiliates' pursuit of R&D and avoiding the dysfunctional consequences of administrative complexity, appear to invite problems of omission and redundancy in the management of their R&D activities; certainly, a substantial number of the decentralized firms exhibited a "looseness" in their control of R&D activities suggesting the possibility of inefficiency. While decentralized firms are more likely than their centralized counterparts to establish R&D abroad, they are also more likely to allow their foreign R&D activities to evolve from technical service activities than they are to create foreign R&D groups through direct placement.

Foreign Government Pressures on Transnational Corporations Regarding R&D

While most of the firms interviewed denied receiving pressure from foreign governments to locate R&D abroad, an appreciable number of foreign laboratories were identified as resulting from host-country government influence. Several conclusions can be drawn from these experiences. Such pressure does exist and can be successful if the conditions under which it is employed are right. The successful influencing of transnational corporations' foreign R&D decisions depends primarily upon the attractiveness of the foreign market: if the foreign market is sufficiently attractive to the firm, then the host-country government can probably induce the firm to locate R&D there. Typically, this means that the developed and advanced developing countries will be most successful in luring the R&D of transnational corporations. Positive inducements, such as tax subsidies, are of little value in influencing the transnational corporation under almost all circumstances. They represent an unnecessary cost to the host government and do not appear to have influenced any R&D location decisions among the firms in our sample.

Negative inducements, such as constraints on profit repatriation and import restrictions, have some influence on TNC decisions to establish local R&D activities. But they work only to the extent that the local scientific and technical infrastructure is sufficiently developed to accommodate and support industrial R&D activities. Unless a country has this infrastructure and the promise of sufficient market volume, the performance of R&D in that country by a transnational corporation will be low level and just sufficient to satisfy government regulations. Furthermore, it would appear that if government regulations increase without an accompanying improvement in local scientific and technical resources available to support additional work, the transnational firm will seriously consider withdrawing from the market. Obviously, negative inducements can only work after a firm has located in a foreign country; once they are imposed to influence the activities of a particular company they undoubtedly discourage future investment activities by other firms not yet in the country.

Benefits Derived by a Host Country from Local R&D by TNCs

Foreign governments anticipate benefits deriving from the transnational corporations pursuit of local R&D activities. If they did not, they would not be motivated to attempting to influence these firms' foreign R&D location decisions. Just what these benefits are, however, is difficult to ascertain. In a sense, the transfer of R&D skills and activities abroad represents the ultimate extension of technology transfer and, as such, is something to be welcomed by foreign countries. There are also somewhat intangible benefits of having R&D localized: technical assistance to government, suppliers, and customers; the general diffusion of capabilities and techniques; and aid in staffing for local universities. What did not exist, however, among the firms in our sample, were conscious, direct efforts by the firms to contribute to the technical capabilities of the host countries.

The passing of R&D-related skills and capabilities from the transnational corporation to the host countries was, by and large, serendipitous. In those few cases of planned, concerted action, the primary interest was in directly supporting the firm's production or marketing systems and not in transferring knowledge or skills. While virtually all of the companies interviewed employed only foreign nationals in their R&D activities abroad, there was very little turnover among the labor force. Furthermore, the turnover that occurred was among transnational corporations and not a diffusion of trained manpower into the local scientific and technical community. Formal training programs were virtually nonexistent, since local individuals employed as R&D staffers were already among the most highly-educated people in their countries. There was also little evidence of the transnational firms' adopting "appropriate" technology to assist the host countries; in fact, most firms expressed little interest in pursuing "appropriate" technology in the future. All in all, the benefits resulting from the local performance of R&D by transnational corporations are rather nonspecific in nature, and tied mostly to whatever upgrading of local institutions the firm feels is necessary in order to improve the efficiency of its production and marketing operations.

IMPLICATIONS FOR POLICYMAKERS

Implications for TNCs

The implications of these conclusions for policymakers in transnational corporations are numerous. R&D can, and is, already being done abroad by other transnational corporations. For the most part, these foreign R&D experiences have been positive ones. There are apparently some real advantages derived from foreign R&D activities that must be compared against the economies resulting from centralized R&D efforts. Among home-market firms, these advantages lie with on-site improvement of technical efficiencies in extractive and assembly operations. For host-market firms, the advantages include being close to the foreign markets, and currying host-country government favor. Among world-market firms, the advantages are in gaining access to foreign scientific communities that would not normally be available to the company.

Market-orientation is not, however, a totally discretionary variable. Economics, product characteristics, tradition, and inertia all conspire along with managerial philosophy to determine a firm's market-orientation. Management style, on the other hand, is a variable much more under management's control. Our study shows that management style is also an important determinant of R&D activity and of inventive and innovative behavior. Basically, the choices are between more or less centralized control over the firm's R&D activities. Companies with a high scientific orientation tend to choose more centralization and thus are better able to exploit and coordinate their worldwide corporate scientific and technical resources. Firms with a greater emphasis on consumer products tend toward low centralization. This allows them to get closer to their markets and to react more quickly to shifts in consumer preferences in those markets. The gain in reaction time is probably obtained at the expense of a loss in product standardization, an increase in R&D redundancy, and a general increase in corporation-wide R&D inefficiency.

A number of companies have acquired foreign R&D operations as a means of establishing R&D abroad. Typically, however, the R&D group is not normally considered in the acquisition planning, and R&D resources are obtained only as riders to the original marketing or financial considerations that initiated the acquisition. In the period immediately following the acquisition, the newly obtained R&D group is usually left alone, with some form of integration gradually occurring over time. Acquisition does, however, represent probably the quickest means of establishing R&D activities abroad, and the experiences reported in this study have generally been positive.

Implications for Potential Host Countries

Conventional wisdom suggests that, to grow technically, countries need some form of scientific and technical infrastructure. Without such a base, technology transferred to a country will wither and die or, at best, stall at a plateau of technical achievement, leaving the country increasingly further behind the technical leaders. A country must have R&D performed locally in order to develop this scientific and technical infrastructure; transnational corporations provide the quickest means of initiating these R&D activities. However, transnational corporations are not all equal in performing R&D, and therein lies the problem for developing countries.

Firms with a host-market orientation, typically involved with the manufacturing and marketing of consumer products, represent the best chance the developing countries have for attracting R&D. These firms need to be close to local markets, they often have decentralized management styles that provide the type of autonomy necessary to allow R&D to be performed abroad, and they have the smallest "critical masses" in their R&D groups, requiring smaller and less sophisticated R&D staffs than their more scientifically oriented counterparts. They do, however, have some drawbacks. The consumer industries are typically not the industries the governments of developing countries want or can afford to encourage. In addition, they tend to rely more on evolution than on direct placement in the establishment of their foreign R&D activities, which is a more time-consuming process. The very autonomy that makes it easy for the host-market companies to establish foreign R&D groups also leaves these groups somewhat isolated from the rest of the firm, thus reducing the local R&D groups' hopes of becoming "plugged into" international scientific and technical communities.

Market potential and scientific and technical infrastructure represent the two most important variables in making a foreign location attractive to a transnational corporation for the pursuit of R&D. The two are somewhat interrelated and so it is difficult, if not impossible, to separate them. It is clear, however, that if a market is a desirable one, and if it is different enough from other markets, it is likely to attract R&D activities without the government resorting to inducements. It is also evident that if the market is not attractive, resorting to inducements will not work, either. Enough transnational firms perceive inadequacies in foreign scientific and technical infrastructures to make that perception a significant barrier to their locating R&D overseas. Thus, not only should a country desiring R&D activities make its market attractive, it should also strive to develop its scientific and technical infrastructure. But those are some of the very reasons why nations want industrial R&D activities in the first place! It would appear that there is a vicious circle operating here that must be addressed by developing country policymakers.

Host-country policymakers should be aware that government pressure has sometimes been successful in influencing the R&D location decisions of the transnational corporations. The key to success, however, has been a desirable market, a scientific and technical infrastructure adequate to support a firm's expansion of R&D activities, and a "captive" firm—one that is already located there. These conditions account for much of the success that the Japanese and French have had in inducing TNCs to locate R&D activities in their countries. Paradoxically, it has been the experience of the firms interviewed that the developed countries have sought to make their locations more attractive, while the developing countries have more typically resorted to negative inducements.

Implications for Parent Countries

Most of the parent countries are also potential host-countries as well. This is because most of the foreign R&D activities of transnational firms are located in developed countries. In this section, however, the considerations discussed are different from those discussed previously, reflecting more the political concerns of "losing" R&D to foreign countries. The impression one gets after talking with members of the R&D community is that there really is very little reason for concern over the establishment of R&D activities abroad. In the first place, most of these R&D activities are low-level, development-oriented activities— not the sort that will threaten the parent country's technological lead. Furthermore, most developed countries are hosting as many R&D activities of foreign firms as their own companies are sending abroad. So, in essence, there is some balance. Also, improvements in the scientific and technical sophistication of foreign markets as a result of hosting the transnational corporations' industrial R&D activities should serve to create new markets for domestic firms.

The matters of concern associated with the establishment of foreign R&D activities by transnational corporations appear to lie with negative reasons at home that drive R&D out. The principal example in our study are the U.S. Food and Drug Administration's R&D regulations that have led to the establishment of a number of R&D activities abroad by firms in the pharmaceutical industry. Any time such negative reasons arise for doing R&D elsewhere, there is reason for concern.

LESSONS FOR FUTURE RESEARCH

Some lessons can be learned from this study relative to methodology, research design, and directions for new research.

Research Methodology

The primary means of data collection for this study consisted of field interviews, which were structured but open-ended. In each, the interviewer carried with him a structured response form. This form, which was annotated by the interviewer during the interview, served primarily to establish some degree of standardization among issues covered. A substantial portion of each session, however, was un-structured and, typically, followed leads that developed from par-ticular responses. This combination of structured interview format and the opportunity for interactive dialogue between the researcher and interviewee allowed the development of considerably richer information than could have been obtained otherwise.

An initial purpose behind the development of the structured interview form was to enable some form of statistical analysis of the data following completion of the sample of interviews. Such a technique had been used in a previous study of R&D management by one of the researchers and had proved then to be of considerable value in adding an empirical dimension to interview data. The focus of the present study, however, was truly exploratory, given the paucity of previous research. Thus, the ability of the researchers to statistically analyze this data was seriously constrained by uncertainty as to the range of responses the hypotheses addressed and by the need to introduce new concepts as the study progressed and our understanding increased.

Given the low level of knowledge about the foreign R&D activities of transnational corporations prior to this study and the exploratory nature of this research, it is premature to expect empirical studies of this subject at present. It is to be hoped, however, that the insights developed by this study will generate the types of hypotheses that can eventually coalesce into a useful, policy-oriented paradigm of the foreign R&D activities of transnational corporations. It is the researchers' opinion that there is still much work to be done before empirical analysis of this subject should be attempted. As a next step in that direction, more structured research into elements of a paradigm of foreign R&D behavior should be encouraged. These studies should initially rely heavily on case and anecdotal evidence with the short-run objective of establishing testable hypotheses.

It should be noted that only two firms refused to participate in our study when approached, and that the average interview time was approximately two and one-half hours, ranging from an hour for the few with little overseas experience to a full day with those with extensive experience. A participation rate as high as this, and a willingness by top R&D managers to devote a substantial amount of time to an academic study, could only be achieved through the use of field interviews. The substantial commitment experienced also attests to current interest in the subject material.

Research Design

The initial research design for this study was to select a sample of firms based upon industry affiliation. As the study progressed, however, insights developed during the interviews and the difficulties involved in coordinating travel schedules resulted in a slightly different sample of firms being selected than was originally anticipated. The sample actually employed covered a wide variety of companies, most of whom had had foreign R&D experience. Given the exploratory nature of this study, this was probably not a serious problem.

In retrospect, the major deficiency in the research design was a failure to anticipate the significance of the market-orientations of the firms in selecting the sample. Any future efforts to continue research into the foreign R&D activities of transnational corporations should pay particular attention to market-orientations and managerial style in selection of the sample of companies to be examined. The former is probably ascertainable from a review of the annual reports issued by the firms; the latter, however, is considerably more difficult to identify outside of an interview situation.

An alternative approach to sample selection would be to examine the R&D activities of transnational corporations in particular foreign markets and to compare the market-orientations and managerial styles of their parent firms.

Suggestions for Further Research

Several areas of additional research have been identified as a result of this study. Perhaps the most important is the relationship between R&D and industrial competitive success. While the findings of the present study allow us to compare R&D behaviors among firms, they do not directly address the ultimate issue of the relationship between R&D behavior and competition. R&D is, after all, of little value by itself. It is the system that encourages R&D and then translates its results into new products or processes that is truly important.

A second important area for research is a more detailed examination of how the R&D programs of transnational corporations are assembled and to what extent the needs and peculiarities of foreign markets are considered. Our study provided some insights into this process, but considerably more work needs to be done if we are to fully understand the role of the transnational corporation in serving various foreign markets.

It would appear that a detailed examination of the process of evolutionary growth in the R&D laboratories of the foreign affiliates of transnational corporations would be a fruitful subject for further research. Most of the foreign laboratories we discussed in the present study experienced some form of evolution. A detailed tracing of significant decisions, resource considerations, and barriers would be most useful.

Given the political nature of the relationship between the transnational corporations and host-country governments, particularly those in the developing world, it would appear that there is considerable merit in examining in greater detail how foreign countries might attract transnational R&D. The present study reports that only in the developed and advanced developing countries were government pressures successful and that, furthermore, in very few cases were positive inducements effective. This would suggest that a more comprehensive inventory of potential government inducements be generated and examined for possible effectiveness.

Similarly, the diffusion of R&D-related skills and activities to host-country scientific and technical communities was found to be serendipitous and less pervasive than might have been anticipated. Further research to identify cases of the effective diffusion of R&D-related skills and capabilities, particularly in developing countries, would be quite useful to transnational corporations and host-country governments alike.

The whole issue of R&D in the developing world needs considerably more attention than has been given it in the past. This might include both case studies of the role that transnational corporations play as members of local scientific and technical communities as well as projective studies seeking to identify effective policies for the development of indigenous R&D resources.

The research reported on in this study suggests that collaborative research, both within and outside the institutional frameworks, will grow in importance in the future. Of special interest to policymakers would be the issues of how to stimulate active R&D support of minority transnational corporate partners in joint-venture activities. Improving the value of technical information-sharing in multilateral research agreements would also be of immediate relevance.

Perhaps the most direct link between R&D and technology transfer in the short run, is the issue of "appropriate technology." A more intensive examination of the role transnational corporations might play in the development and employment of "appropriate" technology is certainly in order. Given the results of this present study, the sources of R&D for "appropriate technology" need to be emphasized in future research.

NOTE

1. Industry Studies Group, Division of Science Resources Studies, U.S. Industrial R&D Spending Abroad," *Reviews of Data on Science Resources*, NSF 79-304, No. 33, April 1979.

Organizational Structures of Foreign R&D Activities Within Transnational Corporations

This appendix presents a sample of organizational diagrams that portray the formal positioning of foreign R&D activities within a number of transnational corporations. The diagrams were selected to reveal a range of possible arrangements and to illustrate a variety of relationships between corporate R&D headquarters and the R&D activities of foreign affiliates. It should be evident from this collection that there is no "one right way" to organizationally structure foreign R&D activities. In fact, nearly every firm interviewed had a structure unique to itself. It should also be evident, however, that many of these structures are variations on the theme of centralized-decentralized management style.

The organizational diagrams that follow are roughly on a continuum with the most centralized designs at the beginning and the least centralized designs at the end. For the most part, these diagrams were constructed from the interview data and are not copied from the firms' formal organizational structure. As such they are more representative of what is really happening within these firms than formal documentation would be.

CENTRALIZED MANAGEMENT OF R&D

The first design (Figure A1) represents a firm exhibiting "absolute centralization" in its management of foreign R&D activities. This corporation is structured on a regional basis for manufacturing and sales, and on a corporate basis for R&D, marketing, and finance. All of the firm's R&D is conducted at corporate headquarters, and is the responsibility of an Executive Vice President who is equal in status to the Executive Vice President in charge of the firm's worldwide manufacturing and sales activities. As is depicted at the bottom of the chart, supervision of research in this corporation is vested in a research committee consisting of the President, the two Executive Vice Presidents, the Director of Research, and a marketing representative. At the same time, each corporate division (Analytical and Process) has a product planning

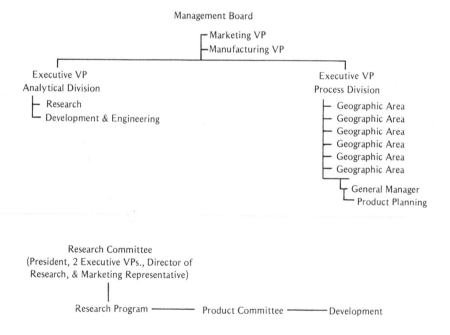

Figure A1. Centralized Management of R&D.

committee that must pass on anything moving out of research into development.

CENTRALIZED R&D
COORDINATION
THROUGH PRODUCT
MANAGERS

The organizational design in Figure A2 shows strong centralized control and coordination over the firm's worldwide R&D activities through use of a global product manager system. This company has four specific product groups, each headed by a Group Vice President. Within each group there are autonomous product lines, with a worldwide product manager for each. One staff member is responsible for the firm's total worldwide R&D portfolio for that particular product. R&D in this company is basically all product development, and the determination of where it will be done is the responsibility of this worldwide engineering manager.

This firm employs a highly sophisticated product planning coordination system which provides, on a near real-time basis, detailed information on the R&D efforts on any product. At present, the firm is moving towards concentration of core product research in

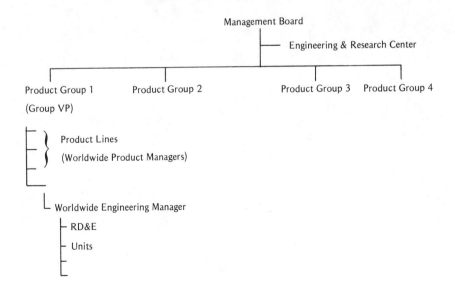

Figure A2. Centralized R&D coordination through product managers.

one particular location, probably the United States, with specific product development support activities being performed at other locations.

CENTRALIZED MANAGEMENT OF FOREIGN R&D WITH MATRIX MANAGEMENT OF DOMESTIC R&D

This particular firm (Figure A3) has adopted a "home-market" orientation. Although it has manufacturing and sales activities abroad, in order to reduce the high costs of exporting its products it perceives R&D as unnecessary for foreign markets because, according to the executive interviewed, its business abroad is "a direct extension of its domestic markets." Its domestic product managers have no line authority over manufacturing, but they do

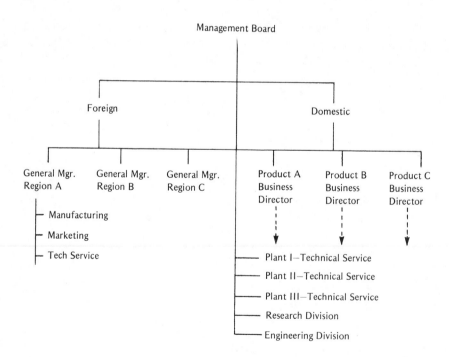

Figure A3. Centralized management of foreign R&D with matrix management of domestic R&D.

have profit and loss responsibility for broad product lines and are responsible for major investment decisions including product development. This is accomplished in a matrix fashion as shown in the diagram.

The foreign operations of this firm are managed by Regional General Managers who, while they have profit center responsibilities, do not have the responsibility for developing businesses and/or product diversification.

COMMITTEE MANAGEMENT OF R&D

The firm in Figure A4 is run by a thirteen-person management board, each member of which has responsibility for either an

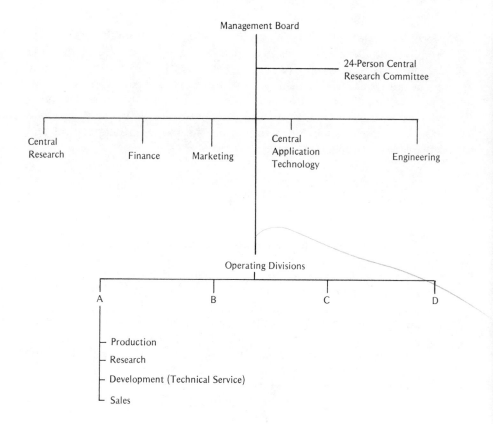

Figure A4. Committee management of R&D.

operating division, a corporate functional division, a region, or two or three central committees. There are ten central committees, including a research committee. Each of the operating divisions is a profit center and is responsible for production, sales, technical service, and a research unit. The Central Application Technology unit is concerned with new processes and new product lines not already in existing divisions.

Within the corporate staff, there is an R&D coordination group (eight people) for administration only. This group coordinates all units within the R&D budget, excluding only the labs directly charged to production (which are usually quality control and technical service labs). Some plant labs may develop new processes and product adaptations, however, and when they get large enough they are included in the R&D budget to reduce "production costs" applied to a particular product.

The central committee is made up of the heads of Central Research, Central Applied Technology, and the Research Unit in the Engineering Division, Research Heads in each of the Operating Divisions, the Divisional Technology Application Directors, and some corporate staff. The central committee meets some four times a year for a thorough review of budgets, coordination problems, documentation regulations, toxicology, publication policy, and program review. It resolves differences among the central research unit and the other divisions.

CENTRALIZED CONTROL
AND COORDINATION

For the company in Figure A5, R&D is a unified function wherein group research is done for all of the affiliates by the central laboratory and is headed up by a Director of Research who reports to a Managing Director sitting on the board. While each functional area within the firm makes inputs into the R&D program for its group, the research group actually makes the decision on the R&D program.

The firm's operating divisions also have R&D capabilities that primarily perform R&D in support of group objectives, but can also bid for R&D projects from the central laboratory. Each operating division laboratory offers its entire R&D program to headquarters for funding and support assessment. As a result of this review, headquarters will provide funds for those activities deemed relevant and useful to other operating companies. Consequently, operating

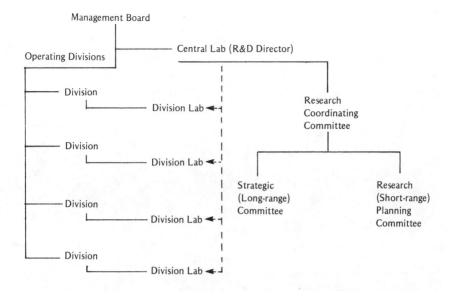

Figure A5. Centralized control and coordination.

division laboratories frequently modify their programs to win "group sponsorship" and thus cut down their own costs.

INDEPENDENT RESEARCH AND DEVELOPMENT UNITS

Research is completely separated from Development in the firm depicted in Figure A6. Each product division is independent within the company and is responsible for product development. There is no separate corporate development responsibility; these efforts are totally the responsibility of the product divisions. Furthermore, there is no formal relationship between the central research group and the divisional development units.

The management structure of this firm is in matrix form, with the twelve Board of Management members each responsible for an operating division. In addition, each country company reports to a board member. The country managers are responsible for exploitation of local opportunities including production, sales, external relations, labor relations, and customer relations. On developmental issues, the country manager is consulted by a division development

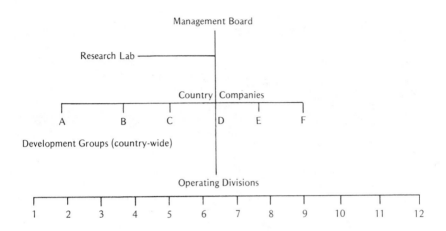

Figure A6. Independent Research and Development units (line responsibility).

manager as to when and where a development lab might be established. Once it is established, the country manager becomes responsible for its efficient operations while the relevant product division determines its R&D program. The determination of the specific R&D projects in these laboratories is made at the divisional level with the development staff having responsibility for worldwide labs in that division.

Research in this firm is supported by an automatic levy against each division, with no accounting to the divisions for the results of these R&D expenditures. To dovetail research interests with development needs, divisional development managers were at one time offered authority over half of the research budget each year, but they decided among themselves that they had no capacity to decide what the research program should be, even for their own divisional objectives.

DOMESTIC PRODUCT LINE MANAGEMENT AND FOREIGN GEOGRAPHIC MANAGEMENT

The R&D management within the firm in Figure A7 is of a hybrid design resulting from a recent decision to move to product line management. This has already been accomplished in the United States, and is being implemented in the European operations. This design is somewhat reminiscent of the design in Figure A3, but this company lacks the same commitment to centralized R&D control.

Currently, the R&D Division has no relationship with the firm's existing businesses, seeking primarily to take the firm into new fields. Corporate policy has traditionally been that the domestic product division managers have had formal product line responsibility worldwide, although this has seldom been exercised. Each division has had its own product-directed R&D activities and, in the division we examined most closely, a technical unit belonging to the division was dedicated to projects with a common interest within the division. Foreign R&D activities were the responsibility

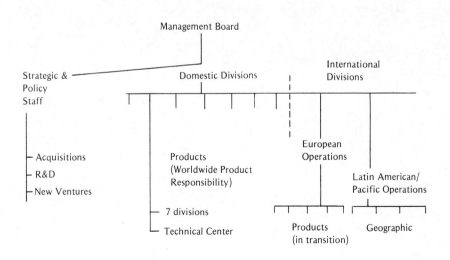

Figure A7. Domestic product line management and foreign geographic management.

of the regional directors and were typically associated with specific operations. These R&D activities were not well coordinated with the firm's domestic efforts.

THREE-LEVEL RESPONSIBILITY

The company in Figure A8 represents a case of corporate growth through merger and the resulting placement of R&D laboratories originally established by the formerly independent firms.

The firm's laboratories report to the manufacturing unit to which they are attached and exist at three levels: branches, departments,

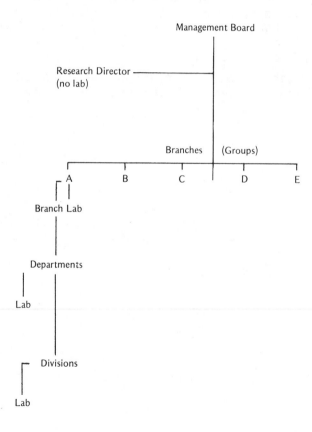

Figure A8. Three-level line responsibility.

and divisions. There are various types of laboratories at each level; they do not report up a line in the laboratory structure, but directly to the operating management at their level. For example, one branch has two large research centers and eight regional laboratories.

Basic research is sponsored by the Central Research Division of the firm, which has no laboratory of its own. Accordingly, it subcontracts the actual R&D to other labs within the company. After leaving the Central Research Division, a project becomes the responsibility of a branch laboratory up to the first stage of development (prototype or application on the factory floor), at which point it is taken over by a department laboratory. The department laboratory may, if large enough, have already done some of the basic research in consultation with the branch laboratory and the science directorate at the corporate level.

Coordination and control of R&D in this firm is accomplished through persuasion and discussion of R&D budgets and by stressing a corporate philosophy of R&D. In the end, however, each business unit has the final say as to what R&D it will pursue.

DECENTRALIZED R&D WITH
CORPORATE SUPERVISION

The firm in Figure A9 has created a position entitled Vice President, Futures, to coordinate new ventures, corporate R&D, and three central corporate laboratories. This individual has dotted-line responsibility for general technical direction over the firm's R&D activities, although R&D program determination and project monitoring are the responsibility of each business unit.

Interaction with the R&D efforts of the business units is effected through a corporate technology committee whose members are drawn from each corporate laboratory and business unit R&D group. This committee governs the sponsoring of work by the business units at the three corporate laboratories and the allocation of that work. The funding of business unit research accounts for about 90 percent of the funding of the corporate laboratories.

LEAD DIVISIONS

Figure A10 shows a corporation where the responsibility for technical direction in particular fields is assigned to lead divisions,

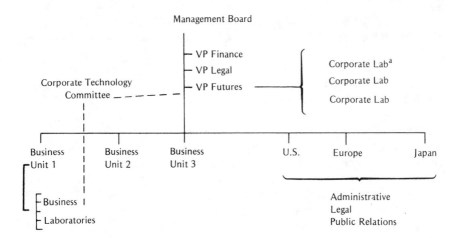

[a]90% of corporation labs' work is funded by the business units through the corporation technology committee.

Figure A9. Decentralized R&D with corporate supervision.

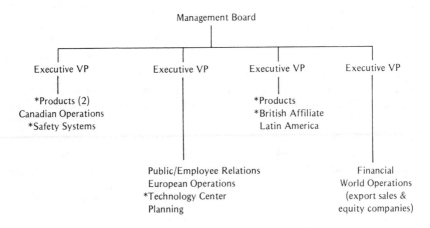

Figure A10. Lead division.

which also supply technology to foreign operations. Coordination of day-to-day technical activities occurs directly between the lead divisions and the foreign affiliates, without involvement of the technology centers, and accounts for about 70 percent of the coordinating effort. In instances where new processes or new materials are involved, however, the technology center becomes involved. The purpose of the technology center, which is also a lead division, is to work on high risk/long-term projects for the corporation as a whole, plus those short-term projects supported by operating divisions and external clients.

The present organizational structure came into existence almost two and one-half years ago. Prior to that the firm had three domestic product groups and an international group. The international group became too large and was reordered into the present structure, which is not the best for technology transfer. In recognition of this, the company has formed steering committees for particular products to link technical efforts between domestic and international divisions to discuss worldwide problems and opportunities—what should be done, and who should do it.

The firm envisions eventually moving towards a product-type organization. This structure poses a problem, however, that worries the firm's management, in that in some countries all of the firm's operations are located in one facility and managed by one individual. If the firm adopts a product-organization, such activities would have to be split, or coordinated differently.

DECENTRALIZED R&D

R&D in the firm in Figure A11 is totally decentralized within the business divisions. The role of the technical director is to review the plans and capabilities of the business divisions and assure top management of their performance. The firm formerly had a centralized R&D laboratory, but it was disbanded four years ago because it was thought to be relatively ineffective.

The technical director has no direct responsibility for the company's R&D activities, although he keeps in close touch with the various laboratory managers and serves as a communication and coordination link between the laboratories. The corporate R&D budget is split among the divisions, with the technical director having a seed-money budget of his own to sponsor R&D of corporate interest in the business unit laboratories.

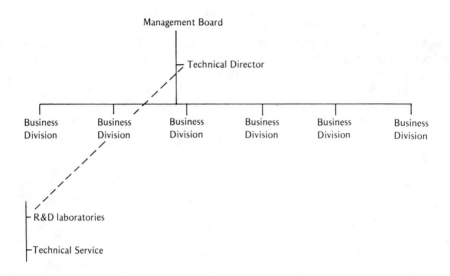

Figure A11. Decentralized R&D.

PART II

The Case Studies

Part II: Three Case Studies

These case studies of some of the R&D activities of Du Pont, Johnson & Johnson, and Unilever were completed during 1977 and 1978 and are, therefore, dated snapshots of the ways in which selected operations were carried out and coordinated. They are much more extensive than the data obtained in Part I, since the companies interviewed there accepted their role in the study only if their time commitment was limited and the results not for attribution. The case studies were also the pilot pieces, in that they were conducted to help guide the remaining interviews with the larger sample. Much of what was described in these studies, therefore, was not followed up in the later interviews with the others.

The companies in the case studies were selected to obtain different patterns of experience but were also determined by their willingness to cooperate. Within the companies cooperating, selections had to be made as to divisions which had experience that would be useful and comprehensible. The Fabrics and Finishes Department of Du Pont was chosen because it provided an adequate characterization of overseas R&D activities and coordination yet was simple enough to describe fairly briefly. Also, the technologies involved are not the most sophisticated, and therefore understandable by a nonchemist, yet they are complex enough to require substantial inputs from a variety of disciplines. The department has several facilities in the United States, some of which the author visited, and several abroad, of which Belgium and Brazil were visited.

Although much more complex, the divisions of Unilever chosen for study were also selected because of their illustration of a variety of aspects of R&D operations and the fact that labs were located in several countries in Europe, the United States, Asia, and Latin America. Labs in England, the United States, Holland, India, and Brazil were visited.

The product area of Johnson & Johnson selected was the main line of business and therefore gave the broadest picture. It was made sufficiently simple for the author to comprehend by the fact of substantial decentralization—markedly different from Unilever.

In each case significant changes have been made in the organization and responsibilities of R&D units since the completion of the studies. However, the cases still provide good illustrations of the ways in which R&D activities are conducted and the problems faced by international companies in coordinating overseas laboratories.

Overseas R&D in Du Pont's Fabrics and Finishes Department

INTRODUCTION

For the Du Pont Company, 1977 was the 175th year of production of a wide range of chemicals. For the first 100 years, Du Pont was exclusively a producer of explosives; over the next 75, it has diversified into virtually every line of industrial chemicals. Its 1700 different product lines fall into four general categories: chemicals, plastics, specialty products, and fibers. The fiber area accounted for about one-third of some $9 billion of total sales in 1977. Commodity chemicals (such as sulphuric acid) and special purpose chemicals (such as dyes and pigments) constitute around 20 percent of total sales. Plastics include elastomers (neoprene), films ("Mylar" polyester), engineering resins ("Zytel" nylon), and others, accounting for another 25 percent of total sales. Specialty products, which include medical products (X-ray films and pharmaceuticals), printing products, electronics, explosives, and agricultural chemicals, accounted for the remainder.

To support these diversified products and to meet the many changes in the market in the United States and around the world, Du Pont has had average annual R&D expenditures of around $350 million during 1975-1977. The purposes of the R&D activities, complemented by an extensive technical staff, are to generate product and processing improvements to achieve a relatively short-term payout and to develop new technology to create a longer-term competitive edge. New product introductions during the 1960s were double those of the 1950s, and were the basis for launching over eighty new ventures of which two-thirds became viable businesses with combined sales in 1966 of $1 billion. During the first part of the 1970s some two-thirds of R&D outlays supported modifications in existing product lines and processes to improve the cost position of the company, with the rest going into exploratory work and new ventures. The necessity to maintain and improve new lines will increase the proportion of research dedicated to modifications to nearly three-fourths of the R&D budget in the next few years. Consequently, exploratory research into new ventures will decline to about one-fourth of total R&D expenditures. This still leaves around $90 million per year for discoveries.

In addition to product and processing improvements and new ventures, R&D expenditures are also directed at reducing raw materials cost through more efficient inputs or use of cheaper materials while maintaining quality. Finally, substantial R&D time is required in regulatory matters and in environmental problems. Scientists and engineers have become involved in technical problems related to compliance with government regulations, as well as with negotiations with regulatory authorities. The preparation for public hearings of testimony about the programs (rising rapidly above $600 million) that Du Pont has committed for facilities to control pollution of air, water, and land also requires considerable R&D support. Although the majority of these expenditures are made within the United States, they are available for support of product development and regulatory affairs in every other country in which the company operates.

ORGANIZATION FOR
R&D ACTIVITIES

Organizationally, Du Pont is divided into two major segments—nine industrial departments and eleven staff departments. The industrial departments are distinguished according to product specialization but include the International Department, which administers all activities outside of the United States. Among the staff departments, the Central Research and Development Department and the Engineering Department are directly related to R,D,&E

activities. Their support of R,D,&E activities overseas goes through the R&D labs of industrial departments, which are basically organized as follows:

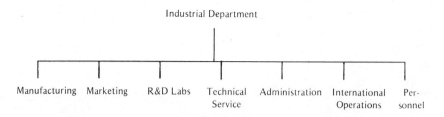

R&D activities within each of the industrial departments are organized differently, and each has its own R&D laboratories at the Du Pont Experimental Station in Wilmington, Delaware.

It is not possible to describe the entire R&D operation of the Du Pont Company—even from an organizational or managerial viewpoint—without writing several volumes. The objective of this study is much more modest: to illustrate how and why R&D labs are established abroad and the ways in which R&D activities overseas are integrated with and supported by those at the center. Direct contact occurs between foreign affiliates and the relevant industrial departments in Wilmington and their manufacturing or R&D activities in the United States. When a foreign affiliate has either an R,D,&E problem or a manufacturing problem, it contacts the appropriate technical person in the Industrial Department, who either puts it in direct contact with professionals in the U.S. labs or manufacturing plants, or goes to one of the staff departments, depending on the type of problem raised. These ties between industrial departments and foreign affiliates are illustrated in detail in the study of the Fabrics and Finishes Department in the next section.

International Department

The International Department has forty-five wholly owned subsidiaries and partly owned affiliates in twenty-five countries around the world, encompassing the one hundred facilities and employing over 30,000 people. Total sales in 1977 were over $2.5 billion, accounting for more than a quarter of the total sales of the company.

The International Department is organized into four major regions and five different staff functions as shown in Figure A1.1.

Although there is liaison with Du Pont of Canada, this is a separate entity managed almost wholly by the Canadian officials. The largest

Figure A1.1. Organization of Du Pont's International Department.

unit in the International Department is the European headquarters at Geneva, with about 1100 personnel. Since two divisions are managed out of Wilmington, the number of personnel can be much smaller by relying on functional staff in the headquarters company. The point of contact in industrial departments will be different according to the department and the type of problem involved. Sometimes major channels of communication develop through the "international marketing" section of an industrial department and at other times through a specially established "international operations" division.

The International Department was formed in 1956 as a profit center responsible for all activities outside the United States. Since then a collaborative responsibility has evolved between International and the various industrial departments. The International Department provides an interface with the responsibility to follow a variety of situations, making sure that all corporate interests are covered. It has no R&D facilities in the United States, but it shares responsibilities for such activities in its affiliates overseas. It is involved in the formulation of agreements between the parent company and an affiliate in the setting up of research facilities, relating to the flow of information and know-how. In practice, the company has *not* developed products de novo abroad that would merely appeal to a specialized market in a few foreign countries. Rather, it has adopted the position of extending abroad what has been successful in U.S. markets. This is not a "high policy" decision, but simply traditional practice within the company. Consequently, it has not sought to develop "intermediate technologies" for application in specific situations in host countries. However, affiliates in particular countries have on occasion developed products for specialized markets without the products having been innovated in the United States. The initiative and justification for the research program have, however, come from the foreign affiliate itself.

Information Systems Department

One of the staff departments that assists in the transmission of R&D data throughout the company is the Information Systems Department. It has two major divisions: the Central Informations Services Division and the Central Computer Systems-Telecommunications Division. The Computer Systems Division has some ten large computers, one of which is in Europe, to store, process, and disseminate data and to facilitate telecommunications. It also assists systems developments within various units of the company and helps their computer operations.

The Central Information Services Division has some library facilities at Wilmington, information resources, a patent and reports service, and a unit providing information on toxic substances. This division is essentially a data-scanning service available to all units of the company, including on R,D,&E topics. Foreign affiliates gain access to this division through the international operations divisions of related industrial departments.

There are three other technical libraries—located in Du Pont's headquarters building, at Chestnut Run, and at Louviers—besides the library at the Experimental Station. The information resources section provides a literature scanning for any technical and commercial (but not financial) subject requested by any department; there are eight professionals in the section, mostly chemical engineers by background. Each department must subscribe to this service, at a prorated annual fee. The International Department subscribes to these services, and its affiliates can draw directly on them.

The patent and reports service section scans all patents relevant to company operations and provides a monthly digest and technical reports. It also keeps internal department reports in abstract form so that it can refer any person to the appropriate industrial departments, which have kept the full research report. The department will then release the report to any other department requesting it. Thus, all research reports are filed by the industrial department or with the information service division when appropriate.

The International Department makes this information available to its subsidiaries. In addition, the information resources section has chemical abstracts and other publicly available information relative to commercialization problems facing company products. It is also tied into the National Technical Information Service in the U.S. Department of Commerce and to private data services such as those provided by Lockheed and Predicast. Almost all its information concerns questions relating to marketing and product development. Data related to more fundamental research is located at the Experimental Station.

To date, there has been little significant use of this department's data by affiliates in developing countries because some have inadequate information about what is available to them, and because the activities of subsidiaries have not yet justified its use.

Engineering Department

The engineering staff has four major divisions, providing R&D, design, engineering, and construction. The R&D division (located at the Experimental Station in Wilmington) develops new equipment for manufacturing or other operations in the various departments and provides consultation on manufacturing processes. Although its activities do not normally extend abroad directly, overseas affiliates do benefit from them since they get a design package, based on this research, fully tested in U.S. manufacturing facilities.

The design division is responsible for converting basic data from industrial departments into specifications for buildings. The construction department is responsible for actual construction (including contracting). Engineering Services is a consultant to industrial departments and other engineering units in the Engineering Department, and has contracted for most company construction in Europe and for major plants in Brazil, Argentina, Mexico, and Iran. When local contractors are used, local procedures are followed in the civil engineering aspects, but established Du Pont procedures are followed in the chemical-process aspects.

Central Research and Development Department

The central research and development activities are located at the Experimental Station in Wilmington. This department is responsible for fundamental research in many types of science, including biochemistry, radiation physics, and physical chemistry. It does not have primary responsibility for toxicology testing, most of which is done at the Haskell Laboratory outside Wilmington. It has "land-lord" responsibility for the Station and for the provision of administrative, medical, and library services to all of the laboratories at the Experimental Station. There are nine departmental R&D labs at the Station, plus that of Central Research and Development. In total, there are more than 4000 professionals at the station including 1000 Ph.D.'s, 500 M.S.'s and B.S.'s, and 2500 technicians. In addition, there are between 150 and 200 engineers with the two engineering labs, which provides staff to all the other units. Among the Ph.D.'s, 700 are in chemistry (organic, physical, and inorganic), with over 50 physicists and 30 biologists; the remainder are spread over a number of other disciplines. Among the 150 Masters are 75 with degrees in chemistry and 75 with chemical engineering degrees. The BS degrees are also divided between chemistry and chemical engineering. Experts exist also in computer sciences, pharmacology, biomedical, and other disciplines. This expertise is supplemented by a number of consultants in various specialities such as biomedical engineering; the biomedical area is probably more closely associated with university activities, than any other department. However, all the labs have periodic research seminars on various esoteric problems, drawing on expertise from universities around the world.

The Experimental Station, as a common location, has allowed numerous communication channels to develop among the various departmental labs. Although scientists tend to group themselves according to their disciplines or interests (so that organic chemists have their seminars, the polymer scientists have theirs, the metallic scientists have still another, etc.), efforts are made to have each professional manager report orally on his work at least once a year so that others can keep up with what is being done throughout the Experimental Station.

Some 170 professionals have project responsibilities at the Station. Their success is evidenced by the publication of over a hundred scientific papers each year, plus those presented at professional meetings and numerous patents and patent applications. Another evidence of their expertise is the large number of papers presented to professional audiences and the frequency of leaves taken to become visiting lecturers or professors at universities in the United States and several countries.

The Experimental Station also has an extensive library containing 1400 periodicals, 40,000 volumes of scientific works, and numerous government documents. Twenty-six employees ensure that these facilities meet the requirements of the various professionals in the labs.

The Central Research and Development Department itself has 190 Ph.D.'s, plus 500 technicians. The Ph.D.'s are predominantly in chemistry, most of which are in physical chemistry, followed by inorganic, analytical, and organic chemistry. In addition, there are a number of physicists and a few chemical engineers. The results of the department's work are written into research reports, which are sent to the relevant industrial departments and to various affiliates overseas on a "need to know" basis. The developments of the department will be picked up by industrial departments, depending on fit with their product line and their view of the innovation's economic feasibility. If an industrial department does not pick it up, the Central Research and Development Department may decide to develop the product further itself, seeking to demonstrate its potential profitability through activities of its own "development division." After further development, it might mount a promotional campaign within the company to induce the relevant industrial department to pick up the product or processes involved.

The facilities available to the scientists in this department and throughout the Experimental Station are the latest available, including some developed at the Station itself. For example, there are several varieties of spectrometers, emission spectrographs, different X-ray units, electron microscopes, gas chromatographs, ultracentrifuges, light scattering equipment, high-energy accelerators, and special equipment for magnetic electrical measurements, to mention only a few. In addition, there are some specialized facilities such as laboratories equipped for carrying out reactions at high pressures and others for growing plants in a controlled environment. These permit types of exploratory research that would not otherwise be feasible. Specialized and analytical laboratories also provide researchers with assistance in development of new methods and techniques of measurement and analysis. (For example, spectroscopy laboratories provide assistance in molecular structure studies of the most sophisticated sort.) A technical facilities division offers equipment and experienced personnel to assist in techniques not normally available in the laboratory; shop facilities are available for repair and development of special laboratory apparatus; and the instrument engineering group provides consultation, design, and development services in the fields of analytical instrumentation, process control, metrology, electronics, and real-time computer hardware-software systems for the scientists at the Station.

All of these facilities are available directly to the industrial departments in support of their R&D activities and indirectly to the overseas affiliates through the industrial departments as needed to assist in overseas R&D activities.

FABRICS AND FINISHES
DEPARTMENT

The product line of the Fabrics and Finishes Department (F&F) comprises a wide range of paints, plastic coatings, adhesives, waxes, industrial fabrics, and consumer products. The business is predominately in auto finishes, refinishes, and industrial and trade paints, with relatively small sales volumes in the rest. The divisions of the Department are as follows:

Finishes (marketing)
Industrial products (marketing)
Consumer products
Manufacturing (finishes and industrial products)
R&D
International operations
Accounting and business analysis
Personnel and industrial relations

Within the finishes division are auto body paints, sold to the auto manufacturers; refinishes for auto body shops; trade paints (interior and exterior house paints, plant and machinery paints for industry, etc.); packaging finishes (coatings on plastic films and "Teflon" nonstick coatings); and industrial coil coatings. Among the industrial products are plastics on cloth, high-performance wire enamels for the aerospace industry, nonstick coatings for cookware, panels for aircraft and aerospace equipment, and a sophisticated synthetic simulating the appearance of marble, but with greater durability. Consumer products include household cements, car waxes and de-icers, and motor additives.

Given that the major sales of F&F are automotive paints, most of the research activity supports these products. Research activity in paints goes back to the early 1920s, when auto painting was done with numerous coats, by hand, with twenty-five days of drying time. Auto production lines were frequently clogged and even halted by delays in the painting process. The industry needed a finish that could be applied, dried, and polished in a few hours and yet could withstand sunlight, texture change, and abuse. "Duco" nitrocellulose lacquer, which met these criteria, was discovered by accident in 1920. Chemists at a Du Pont plant were attempting to remove light streaks in celluloid motion picture film. A new batch of cellulose base formula was put in a barrel and placed outside the plant for later use. Because of an electrical power failure, the research was temporarily delayed and the barrel was forgotten. It sat in the sun for three days, and when recovered was found to consist of a thin, syrupy liquid, rather than a stiff jelly as was expected. Since the chemists had been looking for a lacquer with greater film thickness than existing types, they began research immediately. Three years later they had a product that was applied on the first of many millions of cars to be finished with "Duco". Multiple colors also were made available, expanding consumer demand not only for colors on new cars but also for refinishing. The ability to spray the paint reduced process time to hours and led to the expansion of many repainting shops throughout the country. By the mid 1920s, the paint was being applied to such makes as Cadillac, Buick, Cleveland, Franklin, Lexington, Marmon, and Moon.

World War II sharply curtailed research on finishes, but it was begun again in earnest in the late 1940s. In 1948 Du Point scientists began researching new lacquers to find entirely new coating chemicals; F&F devoted 135 man-years of research and investigation to the project, at a cost of $3 million over five years. The result was "Lucite" acrylic lacquer, put on the market in the 1950s. It had three times the durability of "Duco," with an air-dry finish that could be readily spot-repaired. It became the finish of preference and is today Du Pont's most popular finish.

A new enamel technology was developed to improve "Dulux" alkyd enamel, which had been developed in 1929; the improved product was introduced in 1956, and in 1966 a vastly improved "Dulux" that dried 50 percent faster and had greater durability was introduced to the refinish market. In 1970, "Centari," an acrylic enamel, was announced. A polyurethane enamel, "Imron," was announced in late 1971 for truck fleets and industrial uses. More recently an aircraft finishing system was introduced, based on "Imron." F&F has continued to improve and modify its basic line, supplemented with a line of thinners, reducers, primers, sealers, and companion products for all refinishing operations.

The development of paints requires the combination of four elements—a binder (polymer or resin), pigments (involving color, opacity, and smoothness), a carrier (which evaporates and is in some cases the source of ecological objections to paints), and additives for adjusting balance of properties (flow, corrosion, gloss, etc.). Durability comes from a combination of the binder and the pigments. Considerable research is being done on water-based, high solids, and powder finishes. Each provides a different balance of properties.

The range of paint colors for which pigments have to be found is in the tens of thousands. Each person sees colors differently, particularly from different materials such as paint on wood or metal, as compared to colors from natural light. (In fact, the ability of humans to see colors appears to be an evolutionary process that is still continuing. So not everyone has the same ability to discern a full range of colors, as shown by the phenomena of color blindness or partial color blindness, green-blindness or blue-blindness). There are, therefore, considerable problems in color matching, which was previously done by individual eyesight but is increasingly done electronically. The range of colors becomes narrowed in the market, however, as customers indicate preferences for particular shades, even leading to the possibility of matching them from among different suppliers. Much research, therefore, is directed to matching colors from different suppliers to sell in the refinishing business. Backstopping these activities overseas are F&F's R&D division and international operations division.

R&D Division

The R&D Division of F&F serves all locations and other divisions of the Department. It contains five different activities, each at a different location:

Basic research—Experimental Station, Wilmington
Auto finishes and refinishes, and color technology—Troy, Michigan
Trade and industrial finishes—Marshall Lab, Philadelphia
Production technology—Du Pont manufacturing plants and Marshall Lab
Patents and licenses—Wilmington

Sampling of toxic vapors and dust in the manufacturing processes employed in the plants is done by plant labs, which sometimes send samples and results to the Marshall Lab for checking. These same labs provide technical assistance in manufacturing processes abroad. There is some crossover of research among Du Pont departments when one relies on materials supplied from another. For example, when paints are made from resins supplied by other departments, the research facilites of the two departments get together to check on processes and applications. Sometimes personnel is switched among labs.

The personnel total in F&F's R&D division can be divided as shown in Table A1.1.

The R&D program at Troy is wholly in finishes, and the programs at both the Experimental Station and the Marshall Lab are over 90

Table A1.1. Breakdown of Personnel in F&F's R&D Division, 1978

Location	Professional[a]	Total[b] (including technicians and administrators)
Experimental Station	36	75
Marshall Lab:		
Development group (finishes)	50	135
Development group (nonfinishes)	9	25
Production technology	42	98
Troy Lab	55	180[c]
Total	192	513

[a]Includes B.S., M.S., and Ph.D. degrees
[b]Includes all support and administrative personnel
[c]The large number of technical and administrative personnel occurs because Troy is an independent lab with its own support services.

percent in finishes. (There is a fourth lab at Fairfield that is in industrial products, mainly coated fabrics.) The Marshall Lab was built at a cost of $10 million (replacement cost of $25 million today), and is 50 percent larger than that at Troy, which cost $6 million (with a replacement cost of $10 million today). There is no good cost figure for the lab at the Experimental Station, since it developed out of existing facilities.

The R&D activities of F&F were spread over the department for a period of fifteen years after the department's establishment. Because of the success of the department in the early postwar period, a research lab was set up under a technical director, and developmental labs were established under the marketing division of the department. The latter provided technical services for the customer and new product development from existing techniques. The research labs worked on a variety of new concepts from which new products arose; however, several turned out to be unrelated to the business of the department or Du Point itself, despite its ten years of efforts.

New F&F management in 1970 decided to put the two research and development activities into a new R&D division to be funded by the business division of the department. This latter division set short-term objectives that were mainly oriented to the "bottom line." Marketing, therefore, began calling the shots for most of the R&D division's projects. The management concept was to set up a profit center around marketing, shifting the R&D program to a direct response to demands of the "product managers." The product teams thus set the goals and consequently the R&D commitments. Any disagreements went to higher management within F&F.

By 1974, however, there was a gradual shift to a separate divisional program determined by the F&F Director of R&D, who looked at longer-range objectives and the health of the business as a whole. Top management in the F&F Department and the R&D division decided on the shift and simply announced the new orientation, since some of the product managers within the department probably would have never agreed to such a reorientation. Product managers still have a substantial say in what is done, however, for they must agree with the R&D project managers on work to be undertaken by the R&D division. The directors at the three labs have primarily administrative responsibilities, making certain that the projects are fulfilled and resources are adequate. A triumvirate of project managers at Troy and at Marshall plus technical managers in the business division at F&F put their heads together to make recommendations to the Director of the R&D division on F&F's total R&D program.

Experimental Station Lab. The F&F Lab at the Experimental Station has three separate activities: one is basic research on finishes, a second responds to requests from the F&F business division, and a third consists of joint programs with the Central Research and Development Division. A project for the business division is financed by that division, and the lab director may accommodate it without checking with the Director of the R&D division, depending on the personnel available at the time. Basic and longer-run projects are decided upon by the directors of marketing and R&D. These projects are supported by a charge on the various businesses in the F&F Department and are allocated according to the expected benefits—usually about 80 percent to automotive finishes and refinishing (marketing).

The role of the Experimental Station lab is to fill a "technology bank" with its own research ideas from either inside or outside the company. The labs in Troy, Belgium, and Brazil then pull what they want from this bank. The most important area of research at this lab is the discovery of new approaches to low-solvent content coatings.

One of the most important projects in the lab is the simulation of corrosion and corrosion resistance in order to develop more effective anticorrosion paints. The lab is also working on metallic paints, which require very fine grinding of colored pigments to obtain the desired transparency. Specialized test equipment is needed to test chip resistance of the paints, by throwing gravel at the painted material at various angles.

Joint projects with the Central Research and Development Division are initiated by F&F, and cooperation occurs in stages, or else joint supervision is set up for different stages of the project. These projects are not really joint in the strictest sense, since each unit specializes as much as possible with the facilities and resources available.

The cost of operating a lab ranges between $120,000 to $140,000 per year per professional; with thirty-six professionals (thirty with Ph.D.'s), the total cost runs to $4 million per year. A single lab room for polymer research requires two chemists and two assistants working in a room 500 square feet containing some $6,000 of equipment, plus support facilities in the lab itself and between $200 and $500 per month in supplies. Specialized equipment—such as infrared spectroscopes, weathering machines, gravelo-meters, and spray equipment—is used to test for heat, light, and water. More expensive equipment for radiology and gas-liquid chromometry or mass spectrometry exist in other labs at the Station but can be used by F&F.

The lab does not do work directly for foreign affiliates unless they need the use of specialized equipment for a particular purpose. Otherwise, the overseas labs will generally modify U.S. technology for their own purposes.

Marshall Lab. The Marshall Lab serves all the manufacturing plants in the F&F Department. It is located at Philadelphia, as a result of historical accident. It was begun in 1950-1951 and was placed next to an existing plant—though now it would probably be sited away from any one plant, if the decision could be made again, since it serves many different plants. Some 75 percent of the personnel's time is likely to be spent on activities wholly independent of plant operations; the remaining 25 percent is spent in scale-up and commercialization of the finishes. This latter part represents a smaller amount of the calendar time of the year, but more of the personnel time.

In its early days, the lab's responsibilities included some auto finishes programs which it retained until early 1960s, when they were consolidated at Flint, Michigan, then at Troy (1973), away from plant operations. In 1968-1969 the production technology group was moved into the R&D division of the manufacturing division.

The size of the lab is determined by the ability and willingness of the marketing division to support the projects. When it does not, a program entitled "Divisional Improvement of Established Business" permits the R&D director to use discretionary funds for projects which he considers to be useful in the long run.

Besides the 100 professionals at the lab, consultants from various technical universities are drawn on for one or two day sessions and open conversation with professionals in the lab. Six such consultants are on retainer the year round from universities such as M.I.T. and Caltech. They provide new perceptions, act as a sounding board, and check the experiments of various professionals. In addition, industrial consultants work on process and product manufacturing problems.

To expand lab capabilities, technical liaison is carried out with various university labs providing two-way communication and inputs into the research program from professors at universities such as Lehigh. Such university labs do some work that the departmental lab might not do. (This means, however, that some university work becomes available to other corporate supporters of the university.) A recent cutback in research expenditures has meant a reduction in such outside consultation. The complete loss of such consultants would not be critical, but these ties are highly useful, and they improve the chances of recruiting good professionals for the labs.

The relationship of the lab's work to manufacturing operations and the channels of communication can be seen from the following organizational chart (Figure A1.2) showing lines of responsibility and communication.

The trade and industrial section assists the process engineering section in the scale-up for production, and it assists the manufacturing support section in its resolution of plant problems. The manufacturing support group works directly with the plant technical supervisor, who may have a staff of from two to twenty people. This liaison is carried out directly through the manufacturing division, and not through the R&D director. If equipment or plant processes are involved that require equipment modification, the corporate staff, which does detail design of all equipment and appropriate cost estimates, consults with the engineering department.

The responsibility of the Marshall (or Troy) Lab ends when it and

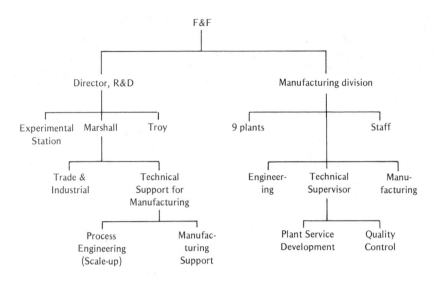

Figure A1.2. Organizational chart of the technical elements of the Fabrics and Finishes Department.

the plant agree that the plant can take over the new process or product. Until then, the costs of development and manufacturing assistance are borne by the R&D division—including the manpower in the plant, raw materials, and overhead, as well as any waste of the product in the learning process. All of these costs come under a "production request," which goes into the R&D budget.

The projects are basically determined by the Marketing Division which sets the marketing objectives for the department, but the technical objectives are set by the R&D division, and therefore projects are cooperatively determined within these two sets of objectives.

The lab may at times "bootleg" some projects on its own, but when it does it risks the decision by Marketing to do nothing with the results. Still, lab officials argue that such projects should be undertaken to go beyond the day-to-day objectives of the Marketing Division.

Projects. The projects of the Marshall Lab extend not only forward to the customer but also backward to the suppliers. For example, some outside companies would like to sell resins to the F&F Department of Du Pont. (In some countries there is a close tie between resin suppliers and research activities in paint companies;

in Spain, for example, Du Pont's affiliate formerly got its paint formulations from a resin supplier.) Some 15,000 raw materials and intermediates are required by the F&F Department, including solvents, pigments, additives, antioxidants, resins, monomers, and synthetic acids. To help determine the suitability of these materials and inputs, which would actually be bought by the Department of Energy and Materials for use by F&F, members of the section on technical support of manufacturing will meet with the suppliers, giving their judgments as to suitability.

The main work of the lab, however, lies in hundreds of different formulations and paint uses requested by customers. Each paint has different properties requiring different experiments to determine durability and stability over the long run. Many different formulations will be required for a single-purpose paint to make certain of success. For example, extremely transparent finishes will change color, requiring a large number of experiments with polymers and different wetting agents (which often destroy the paint properties). Metallic paints require exceptional care to produce uniformity on the tops and sides of automobiles to maintain the same color; too, some metallic paints (clear coated) are not durable under direct exposure to the sun over prolonged periods. Different pigments will be required in aluminum-flake (metallic) lacquer to match the same color in baked enamel, requiring substantial experiments to produce proper modifications.

Other paints must be flexible so that they do not crack or flake when the surface below is bent or flexed. But the color must remain the same as that on metals which are more rigid—and yet be able to flex at temperatures of −20° F. The tradeoffs become more difficult as each paint specification is satisfied. Thus, durability may be lost with an increase in flexibility, or an anticorrosion requirement will be lost. The most difficult part of research is in obtaining the last 10 percent of any set of complex specifications; at that level, one specification is lost as another is gained. The addition of cost constraints makes it even more difficult for the experiments to satisfy all of the requirements.

The project definition may start from zero, such as the definition of a flexible paint for substrates. The specification would involve all aspects including durability, adhesion (of two different substances), and color. The project would start with polymer research on adhesion, putting one or more professional teams on the problem depending on the time pressure. The team might begin by asking, "What do we now know that would help us?" or "Can we use acrylics?" "What can be used for adhesion and still maintain flexibility?" The team might turn to materials from the Troy Lab and the Experimental Station for quick answers to these questions. (In one project on "Dexlar" flexible finishes, the answers were negative in that no single compound would answer all of the questions appropriately, the team then had to look for new (chemistry) leads to make film with new properties.) Two or three prototypes with different properties are then produced, and different colors are made to test against the customer's needs. In the third stage, a larger team is set up under a project manager, dropping the polymer chemist and adding chemists on solvents, additives, etc., to move to the later stages of development.

Once the finish itself is developed, two further stages of research are required—one on the application of the finish to the surfaces to be painted, and the second relative to pollution control.

Research on the method of application sometimes requires experimentation on the mixing of materials at the nozzle of the spray gun. And the finish itself may change in the bake ovens, which use natural gas in paint drying at 250° to 340° F. As gas becomes scarce, automotive paints need lower baking temperatures to be able to use coal-steam heat for drying, which can go to around 200° F. The new finish may be required to permit drying at this temperature. Here again various elements of the process are interlocked, requiring tradeoffs. For example, "reactive finishes" can be dried at 200° F, or below, but to do so, two different chemicals need to be mixed at the nozzle to get the appropriate results. Much experimentation is required to achieve success.

Pollution control research is also required since the solvents in paints go up smokestacks as they dry out. A water base would eliminate this, but a water-based paint requires a more highly controlled environment in production and does not always have the same durability or stability qualities as paints based in other solvents. Research continues, however, on powder paints and water-based paints—especially since two auto plants in California are using water-based paints applied in air-conditioned environments. A major problem is not the temperature inside the drying ovens but keeping the air around them at the proper humidity to allow drying.

One of the contributions made by the lab is the development of new equipment in its own experiments, which potentially has application in the plants of customers. For example, many small manufacturers are using electrocoating, which requires Du Pont to experiment in the process to make sure that the customer is able to keep his plant in working order.

To be able to most effectively help the customers, a complete pilot operation is attached to the lab in a separate building scaled at ratios between 1:2 and 1:10, employing both multipurpose and specialized equipment. A separate building houses the weathering facility, which simulates light, heat, and rain and can reduce two to three months of weathering to between 500 and 2000 hours (20 to 80 days).

A final aspect of the lab's work is to test the qualities of competitor's products through a variety of experiments. This is done through electron bombardment of the paint to get the mass spectra of a compound, which is then researched in the library to learn similarities of its component elements.

The lab maintains a library covering basic chemistry and chemical engineering, plus some physical chemistry covering the subjects of paints from colors to rheology. A number of periodicals are purchased on finishes technology of customers such as auto and aircraft companies. The local library is also available, of course, as well as that at the Experimental Station, and those at the Franklin Institute in Philadelphia and nearby universities.

A computer facility is available for sophisticated technical programs such as paint formulation, modelling of paint color combinations, shades, mixing processes, and cost functions. The computer model has been developed for the testing of scattering and absorption of the paints, which results are stored in the computer for use on all formulations.

Results. The results of the research at the lab fall into three categories: patents, publications, and internal research reports. An average of about a hundred patents a year are filed by F&F researchers. Several may be filed for on the same invention dealing with the composition of matter, the processes, use, and design of the invention. Only a small number of these will be filed for internationally, simply because of the difficulty and cost of doing so. In contrast to the work coming out of the Experimental Station, the results from the Research Division of F&F are not readily publishable. If any researcher wishes to publish his results, he must obtain permission from management to do so, because of the commercial implications. Only about five or six papers are published each year out of the entire F&F Department, including papers at professional meetings— and only one or two of these come from the Marshall Lab. However, management is becoming increasingly open regarding publication, and more can be expected to be forthcoming in the future.

Internal research reports from the lab are sent to the Information Service Department, which computerizes abstracts of the reports.

All F&F research reports up until 1970 are on file with the department itself, and all since then have been microfilmed. There are overlapping areas of research interest among labs that could be drawn upon—dispersing properties, additives, adhesives, and so on. Still, the reports coming out of the Experimental Station lab of F&F are not the basis of the research work at the Marshall Lab. Rather, there is much more overlap and coordination between the Troy and Marshall labs. These two have clearly defined areas of interest and application, dovetailing their responsibilities.

The extent to which other labs draw on the work at the Marshall Lab is evidenced by the search requests coming from them, about eight per month. Some thirteen search requests originate from within the Marshall Lab itself each month, ranging from a straight patent search to a literature search necessary from writing a report. Searches initiated by affiliates abroad average only two or three per year; they are charged for each project, because the lab's budget is so tight.

A process known as "Selected Dissemination of Information," through which profiles of the trade literature are transmitted, gives "early warning" on new data to those likely to request searches on particular subjects. Either abstracts or full articles and documents will be furnished, and the Central Patent Index can be tapped to obtain a specific profile of an invention patented by others or within Du Pont itself. A London Search Service is employed for international patent information.

Foreign Visits. Another function of the Marshall Lab is to provide some technical assistance to personnel of overseas affiliate labs. Marshall Lab professionals make at least one or two visits to a foreign affiliate lab each year. These may be for some extended time to help set up individual projects, or in lab management. The lab in Canada is close geographically, and direct lines of communication exist among the professionals in each product area; visits are frequent. Foreign professionals also visit the Marshall Lab for a few days each, averaging over twenty visits a year. The Marshall Lab takes no initiative to serve the foreign labs in this way; all requests for such cooperation come through the international operations division, and are usually honored.

Troy Lab. The 60,000 square foot lab at Troy houses fifty-five professionals, working wholly on automotive finishes. It is concerned with the "whole balance of properties" in automotive paint and has a pilot plant that duplicates customer needs and conditions.

The majority of the professionals are organic chemists, but others come from the polymer and physical fields. Over 90 percent of their work is applied chemistry, rather than chemical engineering.

The activities of this lab (not visited by the author) closely parallel those at the Marshall Lab. The Troy Lab includes some personnel who have a two-way responsibility between the manufacturing plants and the lab.

The lab is equipped to duplicate the plants' manufacturing processes on a scaled-down size. It thus requires equipment for mixing, break up of pigments, dispersion of pigments, polymerization of resins, spraying of paints, and weathering and wearing—flexibility, bombardment, etc.—to manufacture and test each paint formula.

The lab at Troy—as do the Experimental Station and the Marshall Lab—sends all of its research reports to the Information Systems Department of the company. This department abstracts the reports, computerizes the abstract, and sends the report back to the lab. Affiliates are not provided an index of all reports since there are simply too many of them. Selected indexes are supplied, but in general overseas affiliates simply request information relative to a particular problem or activity, and the Information Service provides them with abstracts. The F&F Department would have to OK the sending of the report itself, which it would probably do when the foreign affiliate is 100 percent owned by Du Pont.

Equipment needs at the labs become quite sophisticated, especially as color matching has progressed from eyeballing by a skilled "shader" to computer modeling. The matching of the paint must be exceedingly precise, since presently different pieces of a single auto are painted at different locations, and they must match on assembly. The expense of such new equipment mushrooms, since any new equipment ordered always requires additional equipment (peripheral and complementary) to accompany it.

Production Technology. Lab personnel at each plant provide liaison with Manufacturing to make certain that new processes are used properly and to assist in any trouble-shooting, if necessary. Each plant has its own technical personnel under a "technical supervisor," who stays on the site but may be working closely with the appropriate lab. For example, the plant at Toledo, Ohio, covers products in all activities under the direction of both the Troy and Marshall Labs. The technical supervisor at Toledo is responsible for both quality control and product development. The latter unit has seven degreed employees and three technicians, who are also available for support of foreign affiliates, working with their labs or manufacturing plants.

The product development unit actually develops specific colors for autos as requested by customers, within limits described by the Troy Lab in its "standard formulating practice" laid down for each type of finish. Within these limits, Toledo develops a customer's color range and specifications for application of the paint and the drying procedures. If Toledo needs assistance, it can get it from the Troy Lab.

International Operations

Besides the R&D division, the other major division of the F&F Department that is concerned with overseas R&D activities is the International Operations Division. International Operations has direct responsibility for all overseas activities falling within the product line of the F&F Department. It is organized into regional sections that have direct responsibility for operations overseas, plus one liaison section under a technical manager, having responsibility for technical operations of all production locations overseas. Three technical coordinators report to the technical manager.

The regional sections are divided between Europe, and Latin America and the Far East. Europe has two major operations—one in Belgium and one in Spain, which was an acquisition. The Belgian plant is a multiproduct operation, though finishes is the biggest single line. The Latin American and Far East area comprises plants in Mexico, Venezuela, Brazil, and Canada.

The technical manager provides the interface between foreign affiliates and U.S. operations (both lab and manufacturing). He has a "transferee" in most of the overseas plants producing finishes, who is assigned there for two to four years for continuous liaison. This individual normally comes out of the R&D personnel in F&F and is, therefore, fully familiar with the processes and techniques as well as the product line. Communication between him and U.S. operations is mainly through the technical manager, who can (because of his position) put pressure on the domestic unit to cooperate with the foreign affiliate, if such pressure is needed. As a result of frequent visits to the foreign operations, the technical manager can also translate their needs into requests that are understandable back in U.S. operations.

The current technical manager had seven years of experience in the Marshall Lab plus several at a plant site and several in the Belgium affiliate itself. He must know where particular kinds of advice can be obtained and how to set up the most effective channels of communication. Once these are established, the foreign transferee often proceeds directly on specific requests. However, performance by the U.S. operating respondent is frequently improved by going through channels to establish priorities and set time schedules and the amount of effort to be allocated. As in most such communications, the "buddy system" also works very effectively.

The characteristics needed to perform liaison effectively include a high degree of technical knowledge and experience and a personality that permits getting along with quite different types of people. The staff in the technical manager's section has had many years of experience. The coordinator for auto finishes and refinishing has been with Du Pont for over thirty years in various labs and marketing positions; the technical coordinator in auto finishes and refinishing has spent twenty-six years in the Troy Lab working on auto programs. The coordinator in industrial and consumer finishes has spent twenty years in the Marshall Lab working on various industrial R&D programs, including three on informational services, patents, and licensing services, and has the added capability of speaking five languages. One of the staff members was recently sent as the technical representative to the Venezuelan affiliate after several years in manufacturing and five years in the development section of the International Operations Division.

This staff spends a large portion of its time overseas carrying information about new developments in the United States and examining the R&D activities in affiliates abroad. The technical manager himself visits Europe and Brazil three or four times each year and others in Latin America once or twice each year. Visits by other members of the staff to all locations increase the frequency to at least four times a year per affiliate. These visits seek to prevent duplication of effort by affiliates abroad, informing them of the total program in the United States and other affiliates. However, some duplication of effort is necessary to meet specific market needs in each area.

Another technique for establishing effective communication has been the institution of biannual international technical managers meetings, the first of which was held in 1976 at the Troy Lab. It was attended by either one or two representatives from each subsidiary's R&D lab abroad (all non-U.S. citizens), plus ten professionals from the Troy Lab, two project managers from the Marshall Lab, and two from its unit providing technical support to manufacturing. The one-week session was spent in reviewing market needs and product and process developments. Each foreign affiliate reviewed what they were doing, with each person presenting some of his work. A manual produced as a result of the conference was sent to all the domestic operations to enable them to communicate more effectively with foreign affiliates. The meeting also permitted professionals to get to know each other personally, paving the way for them to communicate more directly and readily in the future.

Other personnel exchanges may last up to two years. These are similar to management exchanges at the top levels, which are undertaken to provide familiarization with company procedures and contacts with other personnel. When a new unit is established abroad, Wilmington technical personnel are sent there to train the professionals; the foreign affiliate and select personnel there are brought to the United States and Europe on specific assignments in technical service, quality control, lab procedures, and the like. The objective is to minimize the necessity to locate U.S. personnel abroad. Training at the technical levels is done on the job both in the United States and at the local affiliate. There are no regular courses on topics such as "paint chemistry" or "R&D management"; these must be learned through experience on daily assignments. However, outside training is available on a case-by-case basis, as needed by a technician or scientist for the development of his or her personal career. Local management decides when such training is necessary and provides support for tuition costs. Such courses are usually job specific rather than broadly educational.

DU PONT-BELGIUM

The operation in Belgium was begun in 1959 at the initiation of the International Department. The original purpose of the Belgian operation was to manufacture—without modification—paints based on U.S.-developed formulas for subsidiaries of U.S. auto companies. No R&D lab was needed at the site, but a quality control facility and a plant lab for manufacturing support were. However, its marketing group soon realized that European customers would not readily accept the paint line used in the United States. Product adaptations were required, and some new products were needed to meet the competition, who had different paints, while maintaining U.S. quality levels. It was decided to set up a lab for development of finishes. It was opened in 1962, with the primary purposes of color matching and customer technical service. By 1966 the lab had doubled in size and moved into the developmental phase in paints. A 40 percent increase occurred by 1968, at which time "applied research" for new product adaptation had been added in order to match European types of paints and meet their qualities, which were fairly high. By 1977, the lab was four times its original size, with substantial new facilities, though there was no significant change in its basic program.

The need for Du Pont to develop European paint qualities resulted from Du Pont's not being a market leader in Europe; competitors set the product pattern, forcing its R&D to be reactive. The lab has been able to do very little speculative work on filling any gaps in the market; this is not F&F's style of operation.

In the last expansion, the lab enlarged its color formulation capability and added computer facilities for color matching and modeling. (Du Pont had considered developing a single U.S. computer facility for worldwide use, but this was rejected in favor of smaller computers located at site facilities.) The U.S. color programs will be used in the computer, eliminating between two to four man-years of software development. These expansions have *not* been dictated by cost differentials in R&D. The costs of R&D activities in Europe have not yet altered distribution of work among the labs on the continent. Costs in Spain may be lower, but Belgium cannot shift work there because scientists are not available, nor could they be attracted to work at the site of the plant, which is distant from both Madrid and Barcelona.

Geographic location of the lab was readily decided upon—it was simply put next to the existing manufacturing plant. The location of the plant itself was tied to General Motors' already having located in Belgium, near Antwerp; U.S. cars were being sold into Europe through that port, where Opel cars were also being assembled for U.S. shipment. If Du Pont had not been available locally to supply the paints, it would have lost the business. Therefore, if they were starting again, they would still choose this location, which is central to European auto activities and close to GM's affiliate.

Organization and Responsibilities

To understand the functions and roles of the R&D group in Belgium, it is necessary to place it within the organizational structure and the channels of control in the country and Europe as a whole. The organizational responsibilities for Du Pont-Belgium and F&F activities within it are seen in the Figure A1.4.

There are 400 on-site personnel for F&F production, and 60 in other production facilities. There are 79 employees in the Belgian R&D division, of whom 31 are in refinish and "Teflon," 12 are in

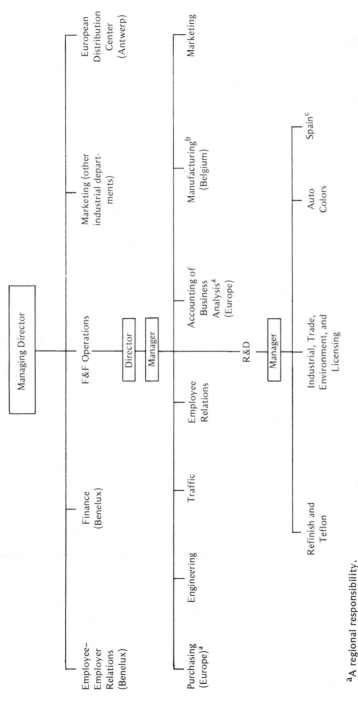

Managing Director

Employee–Employer Relations (Benelux)

Finance (Benelux)

F&F Operations

Director

Manager

Marketing (other industrial departments)

European Distribution Center (Antwerp)

Purchasing (Europe)[a]

Engineering

Traffic

Employee Relations

Accounting of Business Analysis[a] (Europe)

Manufacturing[b] (Belgium)

Marketing

R&D

Manager

Refinish and Teflon

Industrial, Trade, Environment, and Licensing

Auto Colors

Spain[c]

[a]A regional responsibility.

[b]Covers production activities of other Du Pont departments located at same site.

[c]Coordination of other F&F European lab.

Figure A1.3. Organization of Du Pont-Belgium.

industrial, trade, environment, and licensing, and 34 are in auto colors. Twenty-one members of the refinish and "Teflon" staff are in quality control and technical services, while 10 are research chemists. Personnel in the other two areas are also predominantly chemists, with some technicians. Since many staff members have similar backgrounds, and the lab remains fairly small, scientists and technicians can be moved from one group to another fairly readily, providing substantial flexibility.

The personnel are ranked according to ability and experience; a staff chemist should have approximately ten years of experience as a research chemist. A research chemist needs a Ph.D. to enter employment at that level, or may be promoted after approximately five years as a chemist. A chemist must have a university degree or four years at an institute after high school (if training there has been sound). A technician needs a high school diploma and one or two years at a university or institute; he may be promoted to chemist after considerable experience, but only about 10 percent will in fact make the grade. Du Pont has no problem filling these positions, since candidates, are in excess supply; but not all are adequately prepared.

The R&D Manager (Europe) is responsible for all liaison with Marketing (Europe) and with each national company—responding to their needs, getting appropriate technology, helping to institute the latest procedures, maintaining quality control in the plant, and even checking a customer's operating lines to make sure that he is using the product appropriately. It is up to R&D to get a customer's production line moving well; once everything is in order, Marketing takes over—until a production problem arises. R&D then resumes responsibility.

R&D has responded to demand-pull rather than cost-push, since its expansion has tracked fairly constantly with the growth of market sales, remaining at about 3 percent of F&F Europe-wide sales.

As a measure of R&D expansion, the Belgian plant doubled its capacity in 1966. By 1977, production was three times the 1966 level. R&D activities tracked with this expansion, especially as new product lines were added. For example, auto paints in Europe are mostly alkyd-melamine-based, compared to acrylic-based in the United States. The R&D unit has had to expand its capabilities to meet product and consumer needs both for auto paints for new cars and in refinishing. Refinishing is different in each market in Europe, with some using primarily enamels, others using lacquers, etc.

R&D projects follow also the pattern of production in the plant: 70 to 80 percent in auto paints and refinishing, 20 to 25 percent in trade and industrial paints for walls and equipment, and up to 5 percent in adhesives and cements.

Facilities

The facilities of the lab make it largely self-sufficient in serving the company's customers. The projects it has undertaken reflect market demand and its ability to support R&D costs out of revenues.

Within the quality control lab, 21 staff members cover three production shifts. Sixteen provide production quality control and five support plant processes and provide technical service, for example, determining what went wrong in a particular batch rejected by quality control. It is important to reduce the time required in the quality control procedures, since these take between six and twelve hours, compared to 40 hours in mixing a 4000 gallon batch of auto paint.

This unit has automatic spray equipment in a controlled environment (to remove human error in the testing), plus a gas chromatograph, glossmeters, and car analysis equipment similar to that in the United States. A telex facility permits color analysis to be sent to the United States, where it can be corrected to make a specific color as needed. The turnaround time across the ocean for such an analysis can be between ten and twenty minutes, thus facilitating the lab's twenty-four-hour service for the customer, determining what is needed to correct a particular color. Considerable efforts have been made to achieve standardization of raw material supplied for a customer; yet each batch varies even with identical inputs.

The Belgian lab is not yet large enough to effectively conduct fundamental research. Even so, it has relatively more research than operations personnel compared to the same ratios in the United States. There are 58 persons in the Belgian lab, 21 in quality control, and 30 in the Spanish operations, for a total of nearly 110; this should be compared to total operating personnel of 790 in both Spain and Belgium. This ratio of more than one R&D employee to eight production employees is high compared to the United States.

In the research lab itself (outside of the manufacturing unit) a variety of test facilities and equipment permit formulation of paints and testing to meet specifications required by governmental standards and customers. Equipment has been set up for various corrosion tests—mostly against salt spray—since countries have quite different test requirements against corrosion. Weathering tests have been standardized with those in the United States. A list of test equipment includes gravelometers to test against chipping of paints; thermal gradients to check effects of temperature on damage to the paint; equipment to test the physical properties of hardness flexibility and durability under constant temperature and humidity conditions, with bent materials being checked for crazing and cracking; equipment for resin cooking, for testing polyester adhesives, for pigment dispersion, for producing mini-batches of house paint for customer tests, and for industrial paints; equipment to analyze composition of paints; spectrophotometers to determine the color spectrum and permit computer analysis; and an oven which reaches 500°C to test "Teflon." These facilities permit a complete range of testing and mixing of formulations taken from the universe of formulations sent from the United States. The Belgian lab has almost all U.S. formulations on file and can draw on them for customer needs.

The R&D budget in Belgium is twice that of Spain. Half is spent on European paint development, including adaptation of Du Pont technology for European needs. This, of course, is based on Du Pont background technology, the value of which is not calculable.

Lab personnel are given opportunities to improve their expertise by advanced training at technical institutes or universities. Under government regulation, the company gives time off with pay, even if the courses are taken on Saturday or Sunday. Although the company pays the cost of the worker's education and salary while he or she is being educated and the government provides a percentage of the cost as a subsidy to the company, this opportunity is not adequately utilized by personnel, in management's view. This lack of interest may reflect the very small attrition among the ranks of the scientists and technicians, itself reflecting the fact that few employment opportunities are open at this level in Belgium, and personnel hold onto their positions. Over 80 percent of the scientists and technicians are Belgian, mostly from villages near the lab.

Scientists at the lab provide some help to the local universities and technical schools by being "jurists" on degree examinations. In turn, some assistance is provided the lab by professors at these schools, and similar assistance is obtained from other university personnel in the country. These contracts also provide for better success in recruiting additional personnel.

Projects

The scope of the lab's projects is determined by the company's following the lead of European competitors. Not being able to introduce U.S. paint into Europe, Du Pont has had to meet consumer needs as dictated by European company practices. Given the similarity of demand through Europe, the one lab has been able to serve the entire continent, permitting economies of scale in the use of equipment and facilities and in the application of the results of various research projects.

The number of projects depends on available manpower, the supply of which is fairly rigid because of governmental restrictions on layoffs. Unable to reduce the work force readily due to these regulations, the lab will not expand even in good years unless there is a clear need for a continued new level of activity. The determination as to whether to establish "new work" projects rests on whether or not any manpower is available within the existing staff after current market needs are served. About 25 percent of the staff is in "new projects," and the lab could go even higher if pressed by the Marketing Division to do so; but it is not likely to, since it does not want to commit permanent resources to expansion. A few market lines would require only a little development work to break into, but once that was done they would require considerable continuing support from the lab, which it is not now prepared to provide.

The lab's mission is defined as "developing projects for the needs of customers"—with the help of the U.S. labs when needed. European cars are painted predominantly with enamels, though some lacquers are used, similar to predominant U.S. practices. Application of enamels requires three coats over a primer, which is applied electrostatically. Procedures throughout Europe involve several different application conditions, including different methods of spraying (pressures and control) in different climatic conditions. Paints take much abuse in the application process and must be strong enough to withstand it. In addition, in-line repair procedures are more difficult with enamels. Some 20 percent of the autos require some paint repair or touch-up; when repaired in this way, a whole panel must be resprayed. Lacquer can be spot-repaired, since it can be sanded, and a solvent can be used to prepare the surface for repainting. Also, enamels dull more quickly, but they can be polished back into a fine finish—as demanded by the European customer.

Europeans also use fewer metallic paints, and employ a two-coat, metallic enamel, as opposed to U.S. one-coat enamels or lacquers. A clear coat is put on top of the metallic color in Europe to prevent its "graying out" or oxidizing. However, this process does not protect the metallic base against direct sunlight which is not as much a problem in Europe as it is in the United States. The preference for enamels in Europe results from their cheapness and their attractiveness over a short period.

Because of the European preference for enamels, the lab in Belgium has had to develop alkyd-enamels to service the needs of the large dealer-repair shops, as well as independent and small garages. To break into the European market, the lab began with a comparative analysis of European paints. Wilmington sent an experienced polymer chemist to join a team of a Belgian chemist, a Belgian resin chemist, and two technicians. They worked for six months to develop a polymer satisfying European needs. Eighteen months more were required to develop colors for the basic color line, while the polymer was reevaluated in each of the color ranges. Then, when the lab began field testing, it ran into a dulling problem caused by "foul gas" from shop heating systems. They had to modify the formula so that it would perform as satisfactorily in adverse conditions as competitors' paints. The polymer itself proved to be at fault, so the lab had to substantially change the polymer while maintaining other desired qualities.

In addition, the lab had to develop an automotive primer different from that supplied in the United States. An automatic dip tank was used by some customers in applying the primer, so the Belgian lab developed an appropriate process, with technical assistance from Wilmington. A considerable amount of testing was required, since any shutdown of the equipment at the customer's plant would have been disastrous for the company.

The European demand for high quality paints is so great that customers will often pay a proportionately larger premium for paints that have a 10 percent increase in properties such as hardness, flexibility, gloss, durability, appearance, adhesion, resistance to chemicals, and nontoxicity. It is, of course, impossible to get desired levels of *all* of these in a single paint, since to increase one may cause a decrease in another.

To determine which tradeoffs to make and, therefore, how to guide the research projects, marketing managers in each line have individuals checking present and future product demand. Research objectives are formulated from these data by marketing officials and research supervisors, who must determine what is both desired and feasible. Feasibility involves a cost estimate from the very beginning of the project—looking at available raw materials and their cost, and the optimum balance of the properties. However, this optimum balance is seldom reached because of the cost of obtaining each individual property.

The research management for a particular product line meets with marketing managers to determine the lab's ability to fulfill the needs of marketing—the resources needed and the time schedule desired. The schedule is then checked with the technical manager in international operations of the F&F Department. Once the project is approved, running control is the responsibility of the research supervisor, who keeps in touch with the product manager of the marketing division in Belgium.

The determination of the projects to be pursued is more of a rolling system than a formalized decisionmaking structure. Most project work stems from requests for improvements in the product in response to the market's signals and problems, rather than requests for completely new products. The R&D division does not have to accept all such marketing requests; but, if it does, it sets a reasonable schedule and is expected to adhere to it.

Any project may be terminated when it fails certain tests or proves to be too costly in the production process because of the need to use particular materials to achieve the desired qualities. For example, in response to a call for a refinishing product different from anything then available in the Du Pont line, the lab developed and evaluated a polyester putty, which was then tested in the field without success. The project was aborted because the results fell far short of the desired goal, and the goal itself did not have a high priority among marketing objectives.

In another instance, marketing needed a refinishing product for Spain based on nitro-cellulose. They had two products available to start from, but neither met the market needs fully or met Du Pont quality standards. The Spanish competitor had a self-gloss product, while the Du Pont product previously used in the United States for similar purposes required rubbing and polishing. Du Pont-Belgium wanted to enter the market with a high gloss, nitro-synthetic product; it tried to use products existing in Mexico and Brazil, but they were not glossy enough. The lab in Belgium, with the Spanish lab, worked on modifications of technology available throughout the company for over a year. The quality produced was not up to market standards, but market research had found by this time that the product was declining in use. It was estimated that, by the time success was achieved, the size of the market would drop from its former 50 percent of the total refinishing market to about 25 percent—not a market of sufficient size to warrant continued research. The project resources were then shifted to more innovative and useful purposes in support of the Spanish company's objectives. (This example is also illustrative of cooperative research activities. The professionals at the plant lab in Spain spend most of their time helping production adopt Du Pont procedures; the Belgian company has sent two chemists—one from production and one from the lab—to help in these adaptations.)

Project Coordination

Although Belgium needs products that are specific to the European market and different from those in the United States, a substantial amount of overlap and therefore direct assistance flows from the United States into the Belgium lab. For example, in the refinishes market, the Belgian company tried to sell U.S. finishes, which rely on air-dry techniques. European shops baked the paint at 80°C at times, and the customer wanted a paint which could be either air- or heat-dryed. A polymer expert was sent from the Marshall Lab to Belgium to work with the professionals there. He made three different visits over a nine-month period plus two others at later dates. During this time the lab cooked hundreds of batches of polymers and found a balance of ingredients that could be dried both ways. As a result, the product has become one of the leading sales items for the company.

In another instance, U.S. technology sent to the Belgian company had been simply mothballed as not being relevant. The European auto paint division, however, wanted to have the technology applied, so the lab invited a U.S. technician to help transfer it into the lab and thence to manufacturing procedures. The transfer was successful, but the market shifted, and the product was not commercialized.

The extent to which some of the product line depends only on U.S. research and development is exemplified by the adhesives products. Belgium has a specialized line of high-performance adhesives comprising five different products only. These are made to customers' specifications, and the company does not even know what they are used for, since the buyers do not reveal their use. The customer supplies detailed specifications, and Du Pont ships samples back to the customer, who either accepts or rejects. Frequently, the European customers are subsidiaries of Du Pont's U.S. customers, using the same adhesive that the parent company uses. The five products are fairly broad ranged and can be sold for multiple uses, but many of these have to be adapted to specifications set by the customer. These basic five products are right out of the U.S. line, and any technical service required is at the customer's end, changing a solvent or heat-sealing process. If there are major problems, the lab goes back to the U.S. lab for a solution. There is more communication between the Belgian and U.S. labs on this product line than all others put together.

Although many of the European developments are simple modifications of those already in existence in the United States, some are extensive enough to be a new product development, which may or may not be used in the United States or other countries. For example, construction companies were using a staple gun for which they wanted a staple with a finish that would act as a lubricant going into the wood or other material and later as an adhesive, making it hard to get out. The Belgian lab developed such a costing, which might appear to have worldwide applications, but it was not adopted for use in the United States.

In another instance, a coil coating developed in the United States was not commercialized there, and the Swedish marketing group thought it could be used in that country. A U.S. technician was sent to Sweden, where they developed the product successfully and later transferred it to Belgium.

Many of the modifications made by the Belgian lab on U.S. specifications are to help substitute local materials, which are different in quality and reliability. In addition, the materials have different costs, and Belgium wants not only cost reduction but also reliability of sourcing and of quality control by the supplier of the raw material. These problems force a retesting of the materials by the Belgian lab. Materials obviously could be sent to the United States for testing, but it is most desirable to have this done close to the manufacturing and to the customer for quick reactions. Also, it is costly to ship samples of materials, which normally must be done by air to save time. But many airlines will not accept chemicals such as solvents, and ships are too slow.

The use of different materials requires a breaking down of the components of the U.S. product and a substitution of local inputs as close as possible, on a trial basis. Then, careful evaluation by paint chemists with long experience is required to make certain that product quality remains the same. The kind of experience that these chemists have is not available in books, nor is it written up in company reports. The expertise that is frequently needed cannot be obtained even through written communication; telephonic communication is no more adequate. Lab experiments have to be duplicated in order to learn the significance of particular points made by a professional at another lab.

In serving automotive finishes customers, the Troy Lab in the United States is most useful in evaluating pigments and combinations of chemicals suggested by the Belgian lab. The key technology is in formulation (pigment blending) practices. For the Belgian lab to serve GM in Belgium and Vauxhall (GM's subsidiary in Britain) is relatively easy, since they use the same "Lucite" acrylic lacquers as in the United States. However, GM-Opel in Germany uses low-bake synthetic enamels not made by Du Pont-U.S. It would cost Opel too much to change its ovens, so Du Pont-Belgium had to look for numerous alternative resins as binders. It finally found some from a European supplier, eventually building a new capability to supply Opel.

In another instance, a new finish was needed for metal siding on houses, requiring considerable flexibility in the paint. There has been no market in the United States for a paint with such flexibility, so the Belgian lab had to develop specifications in a pioneering effort; it spent two years in development work, with U.S. help in pigment technology. An exchange of three or four people was necessary during this time to conduct the cooperative effort.

The greater part of cooperative activities has developed from the substantial amount of work in the Belgian lab that takes off from prior U.S. research. U.S. labs will attempt to "de-bug" any new product before it gets into the foreign market, but some adaptation is required. Lab officials in Belgium estimated that some 80 percent of their work in refinishing is distinct from that in the United States, while only 50 percent in auto finishes is distinct (except for sales to GM, which are identical with the United States); trade finishes are 100 percent U.S. In addition, most applications procedures are identical with those developed and employed in the United States at some time or other.

Adaptations of U.S. technologies are required to make special formulations for customers. The Belgian lab, therefore, can use U.S. techniques of paint formulation, but it must go through the entire process on its own. For example, the customers of Caterpillar in Europe demanded a higher-gloss paint. Some 75 percent of the lab's work is making such modifications, with 25 percent formulating products for European companies outside of U.S. technologies.

Given that Belgium is developing some new products, technologies could flow back to the United States. Two new "Teflon" nonstick coatings are being examined in the United States, though the needs in that market are a bit different. In the main, however, it is difficult to override the NIH factor, which is strong at the Du Pont headquarters lab. In addition, communication is difficult through long lines, and labs in Du Pont have not been set up for an extensive return flow of technical improvements. One shift that may induce Du Pont to pick up some of the European products is the move to the United States of Volkswagen, which wants paints based on European technology. Conversely, VW may find that conditions in the United States are sufficiently different to make it accept U.S. paints; for example, the United States' location, which is considerably farther south than Europe, means longer periods of heat exposure for autos.

The translation of U.S. formulations and procedures is facilitated by a weekly flow of reports and staff notes from the United States to Belgium. Monthly reviews from all F&F lab locations also provide important insights into current developments. The head of the Belgian lab examines each of these to prevent duplication of work, to apprise Marketing of new product developments, and to check out possible problems in his own product line. Particularly useful is the ability to follow a project failure in the U.S., at the Troy Lab for example, where a research project did not reach commercialization. Belgium once picked up an auto paint line that was developed by U.S. chemists, but not pushed to commercialization. The exchange of chemists among the labs surfaced this idea, which Belgium developed further by seeking the limits of available resins and various combinations of solvents. Belgium examined the properties vis-à-vis customer requirements and ended up with a useful development.

Exchange of personnel and vists of managers and professionals are frequent enough to permit this kind of information exchange. The technical manager at the Belgian lab is at Wilmington or other U.S. labs two or three times a year. Research supervisors have visited the United States three or more times over a ten-year period for two to four weeks at a time, gaining knowledge of new developments and possible new adaptations, meeting people and getting a boost in their motivation to improve research procedures. Such visits are felt to be so desirable that all research supervisors have been to the United States, as have a few of the group leaders, who help maintain contacts and improve communication. The centralization of such contacts through F&F international operations somewhat diminishes the need for person-to-person contacts between labs, but does not eliminate it. Such visits are usually not tied to specific projects, but are for training or personnel exchange purposes. However, visits from Wilmington to Belgium are more frequent than those from Belgium to the United States, since Wilmington chemists frequently go to Europe for professional meetings and visit the Belgian operations en route. In addition, there are usually two "scientists in residence" at Belgium from Wilmington in connection with a specific project.

The telephone expenses of the two top managers in the Belgian lab are $1000 per quarter. The telephone budget of the development manager in international operations in Wilmington is $600 per month, directed primarily at Europe and Latin America. The company cost in this area would be double that, since others call in to Wilmington. The frequency of the use of telexes is four or five times that of the telephone.

Unique Products

Because of distinctions among the European and U.S. markets, Belgium has had to develop some lines unique to Europe, sometimes starting from U.S. technologies or products. Two of these lines have been singularly different from the United States': one was the development of an auto finish for VW, and the other was some new "Teflon" nonstick coatings.

To meet VW's specifications, the lab took an alkyd-melamine type resin and altered it to achieve VW specifications. Du Pont eventually got an order for one color only from the Belgian VW plant, which apparently wanted a local (alternative) supplier of paints and was checking out Du Pont as a potential source. However, the parameters surrounding the paint capabilities were too tight, and it simply would not perform well at all times. Du Pont, therefore, voluntarily backed out of the arrangement with VW until it could develop a technology to support a broader range of products meeting more of the parameters on the paint's use and performance.

In achieving this, they soon ran out of useful U.S. technology and started their own developmental work on European resins, which were not used in the U.S. because they were not tough enough to meet climatic conditions there. European paint companies, however, had continued research in alkyds and synthetic resins, so the Belgian lab had to begin its own independent research on a different range of resins, expanding the lab's facilities and increasing its scientific personnel. (Apparently, Du Pont's anticipation of when Europe might move to acrylics was more accelerated than actually occurred.)

The development of "Teflon" nonstick coatings for the European market was a singular contribution made by the Belgian lab in the non-paint product lines. The company had imported U.S. material, and customers had used it fairly readily, but local manufacture of some types of "Teflon" was begun because of long and slow supply lines and the cost of shipping. Some customers wanted different colors inside the cooking pans, so the lab prepared possible pigment dispersions from U.S. data and FDA approvals on types of pigments in cooking ware. After testing many pigments and finding that some were not stable under high temperatures, they developed a product that was satisfactory to the customer and has remained so for four years without change.

The company has expanded into industrial uses of "Teflon"; for example, it is used on saw blades, where it reduces friction. A French competitor has such coatings, so the Belgian company put chemists on the problem, which required qualities of adhesion to different kinds of metals, durability tests, determinations of thickness and color, and resistance to tree saps and chemicals, etc. The lab had to duplicate all operating conditions for the equipment, or had to meet very detailed specifications of the customer under tests he specified. Eventually, the lab ran tests with the customer on his product to determine whether the coating was really helpful and did, in fact, reduce cost. In other instances, three different types of "Teflon" have been produced and samples sent to the customer for his own long-term testing. The problem of meeting customer needs is difficult because sometimes the customer will not indicate the final use (for secrecy), but still wants specific modifications.

Further problems arise in supplying "Teflon" coatings to a customer because, after deciding through his own tests that he wishes the product, he may demand that the material be supplied regardless of the quantity he wants. If those quantities are small, a high price can result, since the product has a very narrow range of use and will probably be impossible to sell to others. Still, the company may consider that it should be supplied; it will then attempt to negotiate a reasonable price and service fee.

In other instances, industrial uses of "Teflon" had been expanded by the customer's seeing other applications and coming back to the company with a request for uses it had not originally considered. This will cause even more development work to be done by the lab.

To follow up with customers, two individuals have been assigned to Technical Service in "Teflon." They in turn draw on two chemists in the lab assigned to assist in field work with the customer when needed. In order to permit the customer to use the Du Pont seal of quality on their products, quality control officials will test the customer's products along with the customer's own technicians, granting the use of the seal only when the tests are passed satisfactorily.

Coatings of "Teflon" developed for tinplate (a low-cost bakeware) required considerable work at the lab, since tin melts at the temperature normally used for the application of "Teflon." Applications of "Teflon" to electric plates (made of cast metal) for waffle irons, flat toasters, etc., led to problems from the outlining of the heat element on the coating, reactions to hot cheese and grease, blistering on hot spots, and the like. These had to be worked out through laboratory testing. Finally, any new end uses of "Teflon" will raise problems of chemical composition and new toxicity problems.

The Haskell Lab in Delaware tests all new products or assists in toxicity tests. Coated objects are tested at the Marshall Lab first; the Haskell Lab reviews its work and does some of its own testing. (The Haskell Lab is so expensive that the company can afford but one worldwide.) Some toxicity work could be contracted out in Europe, but the criteria for toxic safety are set for all of Du Pont worldwide by the Haskell scientists. To make this work applicable worldwide, a coordinator is located in Geneva whose responsibility is to determine toxicity regulations on each product and feed this information to the affiliates for checking by their marketing divisions to make certain that all requirements have been met on products sold.

To meet multiple national regulatory requirements, which are the responsibility of each local affiliate, the Belgian Lab has put one man full-time on regulatory matters, with the assistance of two technicians. They help each local affiliate write explanations and specifications in use, and make necessary modifications to meet regulatory standards. When help is needed, it is available from the Wilmington labs, though U.S. regulatory standards are frequently different and would not be appropriate for the materials obtained from European sources.

The European Community is attempting to standardize regulations on toxicity, fire hazards in transportation, corrosiveness, etc. To assist in developing and meeting these standards, one chemist at the lab is assigned permanently (20 percent of his time) to work with the European Paint Manufacturers Association, which presents industry problems to the European Community. In addition, the director of the lab spends about 5 percent of his time on regulatory matters. Work in the lab is aimed at eliminating flammability and toxicity by eliminating certain components or carcinogens. In some cases, Du Pont scientists have suggested that certain chemicals are potential carcinogens, but other companies in Europe have not agreed, so Du Pont has had to keep quiet. Du Pont considers itself generally ahead of governments in environmental protection and safety, so not much of the work done in the labs in these areas is occasioned by government regulations themselves. Rather, Du Pont has already anticipated these moves and done the necessary research on its own.

Future Growth of the Lab

The lab's work is essentially in development work and not in research as defined by Du Pont, much less in basic research. To get into the "R" aspect of R&D, the lab would have to have many more Ph.D.'s and more sophisticated equipment than it now has. Even if it could get the Ph.D.'s it would have to train them for several years in paint science and technology. Ph.D.'s in Europe have very little industrial practice, though they are oriented to process engineering and practical applications, at least in Belgium.

One of the lab's first moves might be into polymer chemistry, with two or three Ph.D.'s plus two or three technicians and some expensive equipment. To move into research, however, would mean a 50- to 100-percent jump in the annual budget, which does not make any sense, given the resources in the United States being directed at waterbased paints and other finishes that will be of interest to European markets as well.

What the lab can anticipate eventually is closer ties with Wilmington as its own activities become more sophisticated. At present, specialization between Europe and the United States and coordination through joint projects is beginning at the paint formulation and development levels. It will take five years or more to get to the point where the lab's size, sophistication, and performance is adequate to be melded more closely into the total R&D program of the company.

Application of Experience to
Developing Countries

Lab officials were queried as to how they would interpret their own growth experience in terms of its usefulness as a guide for policy in R&D growth in developing countries. They responded by describing a sequence of events, supposing that a manufacturing plant were developed for auto finishes in a country in South Asia or Latin America. The first need would be a small technical service facility and a quality control lab. The program would involve customer service, testing of products from the line, examination of raw materials inputs, and searching out and determining the suitability of the suppliers of items such as chrome pigments. If these could not be imported because of governmental restrictions, then local supplies would be needed, but these are generally unsatisfactory, requiring adaptations.

The next step would be to move to custom formulation of paint, responding to local climate conditions, different application methods, durability requirements, etc. Manpower limitations would probably cause the major part of the work to be done at the headquarters lab (Belgian or U.S.), with only two or three scientists initially at the lab in the developing country. It would be undesirable to try to develop a full-fledged lab in a developing country principally because of the lack of adequate communication, either by telephone or telex. For example, Iran is spending millions of dollars to convert telexes to Pharsee, which requires an alphabet completely different from Latin, adding significant translation problems. These cannot be circumvented by the telephone, since Europe can call Tehran during only a few hours a day, and calls are often delayed for days.

In addition, R&D facilities should reach a minimum size in order to justify the equipment necessary for testing and development work. A minimum number of fifteen to twenty staff members is required in a quality control and technical service unit, but a development lab requires at least fifty to sixty for effective work and efficient use of equipment. (A lab in Sweden had only fifteen employees and could not justify purchasing the necessary equipment.)

There is a limit on the growth of labs in a developing country. For example, for the lab in Belgium to grow to a 300-employee level would require sales at four times the present level. Even if the lab were ordered by F&F to increase its staff to 200 or 250, it would take at least seven or eight years; the lab simply cannot train more than 25 percent additional staff in any given year.

Training proceeds very much as follows: For the first six months a new scientist is at the bench as a technician, where he is taught procedures and his mistakes are corrected. This process requires the expenditure of about one-third of a professional's time for the first year. Over three years, a half-year of a professional's time is devoted to training one newcomer. Therefore, the first year is virtually a wash: the new professional is hardly one-third as productive as his teacher, who is one-third less productive. Over two years, the new professional would contribute a bit more than was expended on his training; it is only in his third year that he begins to make a strong net contribution. Therefore, if the lab brings in twelve new scientists, over three years it will lose six man-years in teaching and gain the effective equivalent of twenty-one man-years (3 + 6 + 12), with a net gain of fifteen. The training of a professional visiting from another affiliate is a net loss to the training unit, though obviously the affiliate gains and the company as a whole may gain.

In attempting to assess the growth rate of a lab, certain assumptions have to be made: there is no loss of scientists during the process; there are no jumps in the level of business (sales) faster than the lab can develop; there is no limit on the budget; and professional talents are readily available, including computer programming. Given these assumptions, the growth rate of a lab at 20 percent of total personnel added every other year (with the even year used to catch up in the training process) would show something as follows: starting with 50 professionals in the lab, 10 could be added in the first year. These would produce a net output equivalent to 12.5 man-years over a three-year period. The second year would show 60 professionals, with no new ones added, the third year starts with 60, plus another 12 (20 percent) added, making 72 in the fourth year. Five more biannual increments of 15, 17, 21, 25, and 36 would be needed to quadruple the professional manpower from 50 to 215—for a total of 15 years.

Any jumps faster than this would have to be made by hiring either talent already trained by competing companies or advanced Ph.D.'s from independent institutes or labs. These are not readily available in developing countries, so growth rates are probably less than estimated in the training sequence above. Therefore, one can anticipate that it will take maybe ten years for the first doubling over the initial 50 professionals and *another* ten for the second doubling. At the end of this 20-year period, lab capabilities are still within the development phase, and another five or more years would be required to reach the phase of applied research.

Twelve staff members were added at the Belgian lab during 1976-1977, equal to 20 percent of the professionals and technicians in the lab at the time. Of these twelve, four were scientists (one from a competing company), six were technicians, and two were hired for color work, coming out of manufacturing. Despite the prior experience of some, a two-year training period was still required. Absorption of this number of people has been difficult, which shows why it takes twenty years or more to go from simple quality control and technical service through development into applied research projects.

The value of the R&D assistance from Wilmington is reflected in the cost that would have had to be borne in getting to the same point through indigenous resources. If the Belgian lab had had to rely on its own capabilities in developing all paint formulations, it would have delayed the introduction of the product line, would have required "head-hunting" to obtain professionals from competing companies, and costly mistakes would have been made that would have damaged the marketing end of the business. In addition, it would have taken four or five years to shake out the problems and shape up the team of professionals, who would probably have been prima donnas from other companies (even if they could have been attracted) and would still need some years to gain technical capacity equal to that of U.S. professionals. Thus, the lab officials calculated a seven- to eight-year period to absorb the background information that was readily available from the U.S. labs. It did in fact take the Du Pont lab in Belgium between five and six years to achieve the technology levels of their European competitors, even though it had substantial support from Du Pont in the United States.

POLIDURA-BRAZIL

Du Pont purchased a controlling interest in Polidura in 1972, with 100-percent acquisition completed in 1975. The previous owners managed the company until 1975; since then, Du Pont has revamped operations and raised the level of management expertise, especially at the middle levels. It is only in these last years that it has reshaped the R&D activities of the company.

Polidura began in 1956 as a family-owned and operated company. The company had little competition in the automotive finishes area, until 1972. It also produced house paint, wood finishes, and several "speciality" items, some based on German technology. It began production of industrial paints in 1968-1969. In 1969, there were six chemists in the company—two in management, two in the manufacturing plant, and two in a technical lab. In 1970 and 1971, the company began to expand its research on different paint formulations, collecting as much information as possible from all sectors of the industry. It built up 2000 paint formulations, eventually computer-recorded. It also constructed a new lab and began to organize an R&D division. The quality control function already existed in the plant for production lines, and there was a small unit in the lab on control of raw materials. In the early 1970s there were twenty people in the lab, consisting of three professionals (with university degrees), seven technicians, and ten auxiliary personnel.

By 1972, competition had arisen in several of the product lines, particularly automotive finishes, and the company's profitability dropped significantly. Polidura could not keep up with the competition without considerably better internal administration and technical information. For this reason it turned to an outsider with both the management and technical capability to take over the company.

Although Du Pont purchased the company in 1972, it left the former management in charge since a three-year economic boom increased the profitability of the company. There was, consequently, a minimum transfer of technology and minimum interference in the orientation or structure of the company. No effort was made to shift the product lines, which included polyester resins and wood finishes, and were somewhat different from those of Du Pont itself. R&D activities reflected the capabilities of the professionals in the development group, rather than the specialization of the company in particular product lines.

Du Pont took over the reins in 1975 as the boom ended. One of the initial shifts was to reorient lab activities toward the market and the new objectives for the company, modified by the capabilities of the professionals in the lab. Given that Du Pont had no expertise in wood finishes and Polidura's technology in that area came from a German company, nothing was done to change this line. The gaps in technical expertise were in the other product lines, where Polidura had no sophisticated development capabilities—none had been needed, since the market would take *anything* up to 1975, including any paint quality or color, even on cars, since customers were lined up to buy them. To meet the market shifts after 1975, Du Pont brought in not only new products and process technology but also R&D management and professional discipline—which had been provided previously by the owner of the company on a personal basis. F&F withholds nothing on the technical aspects of production or product specifications on the line that has been agreed to for Brazil, but the Brazilian affiliate cannot go into just any product line it wishes. It would have to persuade Du Pont in Wilminton of the advisability of entering the market with new products.

Organization

Since Du Pont was already in Brazil in other lines in 1972, Polidura was purchased by and made a division of Du Pont do Brasil. In addition to Polidura, the company includes divisions covering textile fibers, photo products, explosives, ag-chemicals, elastomers, and other industrial products.

The Polidura division had been organized with a sales manager for each product line, plus a plant manager for all manufacturing, and an administrative services manager. The developmental and technical services activities of the division have been reorganized to directly support the various marketing and manufacturing activities.

An R&D lab comprises research groups in resins and polymers, one in primers and wood finishes, one in water-base/electrocoating and trades paints, and one in automotive top coats. A dispersion lab handles process engineering; a plant-support lab is concerned primarily with quality control and making adjustments on imperfect batches; a materials lab does analysis of inputs, which are of critical importance, given the necessity to use local materials different from those normally employed by Du Pont; and there is an automotive lab, and marketing-support group lab that provides technical service for troubleshooting with customers.

This new organizational structure is not yet fully manned, but it demonstrates to existing professionals the orientation which is expected of them, focusing on the support of the specific product lines. None of the product lines has been changed, but specific items within each have been altered through elimination of some products and the addition of others. Although there has been an increase in R&D effort on the refinishes business, it largely tracks with the importance of the different product lines. Table A1.2 shows percentages of total sales by volume (not by value) during 1972 and 1976.

Table A1.2. Percent of Total Sales by Volume of Polidura-Brazil Products, 1972 and 1976

Products	1972	1976
Industrial paints	25%	20%
Trade paints	33	27
Wood Finishes	15	24
Auto[a]	11	8
Refinishes	8	13
Resins[b]	8	7

[a]The price of auto paints is at least twice that per gallon in the trade or industrial paints; therefore, in value terms the sales in auto paints would be greater than any of the product lines except "trade".

[b]Resins are really plastics for industrial manufacutring of buttons and similar items and are not really a finish at all.

The profitability of the paint industry is not great because all of its products are under governmental price controls. However, new products are not under the same price controls as old products, and therefore an effort to introduce new ones pays off both in negotiations with the government, which is interested in industrial expansion, and in advertising. Labs are thus being oriented to new product development, but also have to spend considerable time on the substitution of local raw materials for what would otherwise be imported. Imports of materials are down from about 40 percent of total content to around 20 percent, a reduction which has been hastened by high duties and a 360-day prior deposit for all imports. At an effective cost of money of 18 to 20 percent, to leave this amount in a deposit with the government, not earning interest, substantially increases the cost of imports.

In 1972 the paint industry was importing most of its raw material but needed to substitute local sources because of the high duties and delays in the receipt of materials. This substitution was made somewhat easier because paints in Brazil were different from those either in the United States or Europe, due to different specifications and application equipment; too, metallic paints are not widely used in Brazil since they require an imported acrylic resin. Only about 20 percent of the paints used are metallic even now, with the rest being solid and mostly alkyd-melamine enamels for autos. VW, which has 60 percent of the Brazilian passenger car market, has used the same types of paints as in Europe—metallic paints are on only three percent of its cars. Polidura had drawn its technical assistance in alkyd-melamine enamels from a German company that supplied VW there.

In separating out the functions of the lab, Du Pont found that Polidura had mixed up the cost of research, development, and technical services with total company cost so that they were virtually incalculable. In Du Pont practice, materials labs and quality control activities are a part of manufacturing cost, while developmental activities frequently are charged to marketing. Du Pont officials apparently see a need for a development lab oriented toward local market needs, and for the inflow of non-Du Pont technology, covering products that will be developed in Brazil but not necessarily out of the Du Pont line. The company changed the orientation of different lab functions to daily sales, development efforts, and longer-term market research support. Sales-support effort is strictly in the form of technical service to meet customer needs; new development efforts would be aimed at new colors and new paints; and marketing support will look at the use of new paints over the long term. A shift occurs among these groups according to what is happening in the market. The development and sales groups grow as final sales drop, in an effort to increase sales and maintain company revenues. When sales rise, more effort is put into direct manufacturing and an examination of the longer-term growth of the company.

Facilities and Staffing

The entire laboratory support of manufacturing includes 106 professionals and auxiliary personnel. The thirty-six professionals include ten in the development lab, twelve in sales support (formulation of samples for customers), five in technical services for marketing, five in the materials lab, and four in quality control. By far the largest portion of the auxiliary personnel are in quality control and sales support; five or six of these have university backgrounds, with the rest mostly high school graduates. Of the professional group, only five or six have a Ph.D. (in chemistry or engineering). There is a smaller ratio of professionals (few of whom equal U.S. training or capabilities) among the staff. The educational level of the entire staff is lower; much on-the-job training is necessary in order to bring the group to effective levels of performance.

The lab facilities at present are not up to U.S. standards. Among the paint competitors in Brazil, the only ones currently with major R&D labs are Hoechst and Bunge-Borne, though the Brazilian company, Ideal, intends to build a substantial lab in the southern part of the country.

The Polidura lab comprises three separate facilities: a building of 1400 square meters in the main location; another equipped for production modelling and testing; and another for quality control, located next to the plant. Presently, there is a small chemical testing laboratory, which is rudimentary, having only two chemists and two technicians. A large laboratory provides sales support through multiple formulations of paints, though the work done here is largely of a repetitive nature. Only a small lab supports refinishes, despite the substantial sales in this area. Equipment in the lab duplicates the application methods of customers, but there is as yet no electronic testing equipment. A million-dollar construction and improvement program for these facilities was approved in 1977.

Training of the personnel extends from management through technical levels into safety training. The company has courses on management-by-objectives and problem analysis, but these are not made more specific for application to research management. Nor is there any special program or technical training to improve the scientific expertise of professionals. Technical training is done on-the-job at the plant. However, all technical personnel and professionals take courses on security, quality control organization, and management principles. To date there has been no training of Polidura's professionals abroad in other affiliates of Du Pont, though the company is discussing ways of doing this in Belgium, for example. One Polidura staffer is in the United States on a special program, and several have visited Wilmington. One also took a course in Du Pont (U.S.) in recent years. Nine Polidura professionals have taken courses in São Paulo offered by the association of paint manufacturers. To assist in on-the-job training and to hasten the efficient operation of the development facilities, Du Pont has sent three different U.S. professionals for one to two year assignments in Polidura: a process engineer for one year, an assistant general manager for two years, and a second assistant general manager for a second two-year term.

Du Pont has made special and quite significant efforts to improve safety training throughout the company; Polidura draws heavily on these safety programs, and even sells them to other chemical companies in Brazil (including affiliates of both U.S. and European companies) and to the state-owned petroleum company (Petrobras). Graduates of professional schools come to the company with no safety training at all. R&D personnel are expected to set an example of extreme care in safety matters. The emphasis of Du Pont on safety extends into the safe transport of hazardous chemicals and paints and into the reduction of dangerous materials used in the paint and even by customers. Safety orientation has spread into both suppliers and customers, who visit the plant and are visited by Du Pont personnel. In each case, safety problems are noted and corrective measures are discussed.

Du Pont's concern for safety even extends into its workers' daily lives. The company has taught its Brazilian employees a defensive driving program (developed by the U.S. National Safety Council) it imported from the United States, and has used the same course to train 650 workers in Brazilian companies unrelated to its own operation, as well as to 350 workers in customer companies.

Contrary to U.S. conditions, the universities in Brazil are not closely related to industrial engineering and research, remaining highly theoretical in their approach. Therefore, professionals and technicians trained in the universities and institutes are really "only ready to be shaped." The company is seldom invited to give lectures at the universities or technical institutes, though it feels an obligation to create such ties if they would be well received. Some rather informal ties are being developed between Du Pont personnel and their former professors at these schools. As a consequence, they have two or three students per year working on projects to obtain some industrial experience. However, the kinds of ties that exist in the U.S. are several years away.

Mission of the Labs

The assignment of the labs is to develop "new" products for the company and technical support for marketing and sales. In the first category fall the search for additional products or product lines, the substitution of local raw materials, and formulation changes to meet customer needs in paint qualities and application processes. These processes are mostly spray painting, electrostatic application, and pressurized guns for cars, all of which are oven-cured.

Technical support is provided in both the areas of formulation and processes, suiting them to the structure and requirements of the manufacturing plant to maintain quality. A process engineering section is yet to be formalized, though a small task force of an exchange official from Du Pont and a student engineer and technician is now developing this function.

Since previously almost everyone in the company reported directly to the owner/manager, who made all the decisions, a reorganization of the lab functions has been required—literally starting from scratch. Job and task descriptions need to be written, structure reorganized and put into place, and channels of interaction created among the various parts, including those with Wilmington. The previous management structure was particularistic; some 700 people throughout the company had direct access to the president. The company could not function any longer in this fashion; more modern management techniques had to be put into place.

Lab management techniques require shifting manpower from one unit to another (for example, into process engineering); a reorientation of professionals' thinking toward marketing requirements; the streamlining of the more than 2000 formulas so that they are recorded systematically and are more readily available to researchers; the processing of raw materials to improve quality and standardization—both local (frequently from vegetable oils) and imported—such as TIO_2 and other pigments and solvents. Before the Du Pont purchase, product lines were set by a technical group composed of professionals hired away from other companies. Each used his own background of techniques and formulations, so that there was no conformity and no effort to set up standard procedures or formulas. A high turnover of personnel accentuated this diversity.

The lab is now seeking to train its own personnel and to promote from within, developing uniform processes and formulations and test procedures, putting as many on the computer as possible so that they will be quickly available to any new personnel joining the lab.

Even test methods in the past were not uniform, but ad hoc depending on the experience of each professional; these now require standardization. The manufacturing processes within the company were also diverse, despite the similarity in paints. In some cases they were using methods that have long been rejected by Du Pont. In addition, raw material specifications were deficient and testing was poor. There was an absence of alternative sources of supply of materials so that scheduling was possible or quality could be improved; Du Pont has sought multiple local sourcing, even permitting the purchase of TIO_2 from a German supplier rather than from Du Pont-U.S., which is also a manufacturer of this material.

This transition to new activities and orientations in the lab began two and one-half years ago, after the 100 percent purchase of the company and the departure of the former owner. According to the director of the lab, who has been with the company since 1969, the lab functions are infinitely better organized, relying on Du Pont's experience in paint and research management. It has required nearly two years merely to form the present organization, without even expanding greatly the lab's developmental and technical functions. In order to prepare for these changes, the lab director spent a month in Wilmington with the technical directors in other Du Pont labs in the F&F Department. In addition, the technical manager of International Operations of F&F visits every three months, offering ideas for Polidura to consider.

The specific missions of the lab can be broken down according to the product lines as follows:

Resins and Polymers. The activities of the company here need to be expanded, since polymers are the heart of paints. The lab will rely on Wilmington technology as well as their own efforts in adapting technology to local materials. It will need more chemists in this area, and can hire them out of college, but they will require substantial training: Ph.D.'s are simply not available in the area of alkyd-resins.

Primers. These will require constant effort to increase protection against corrosion from both air and water. Development of a water-based primer will be a new field to Polidura, based on Wilmington research. In addition, the lab will work on a "high-build primer," which can be spread thickly in one application, rather than more thinly in several; this would require stability in the paint, prevention of sagging, and more even application. In addition, research in fillers and additives will be required.

Wood Finishes. This area is still growing fast in Brazil, and Polidura is the leader. Technology is based on European practice, since the furniture industry in Brazil is similar to that in Europe, where standards for semigloss, resin-protected finishes are different from those in the United States. Still, the technology will need updating, and a new license source will probably have to be sought, since the previous one was from ICI; their competitive positions worldwide makes it difficult for Du Pont to remain the licensee, even through an affiliate. Polidura will need what is being done in Europe, but it can then probably duplicate and keep up its progress through Wilmington's basic research.

Water-based Paints. These are essentially related to primers. The lab will be working on top coats of water-based paints out of U.S. know-how, but a greater effort will be required in primers to cut fire hazard in immersion application and to reduce pollution in the evaporation of the solvent.

Auto Top Coats. The auto companies set the color standards, which must be met by Polidura. More work is needed on color styling for the expanding Brazilian market.

Acrylic Technology. Polidura will eventually adopt Du Pont's along with the color/clear-coat system. Polidura will have to adjust Wilmington formulations to local conditions of application, assembly lines, customer preferences, etc., and to the processes of the manufacturing affiliates of European and U.S. companies in Brazil—which are not as up-to-date as their parent companies'. In addition, the different companies use different processes for application of primers and top coats.

Projects and Coordination

The program of R&D projects for Polidura comes directly from an assessment of the market demand, which leads either to adaptation of whatever paints are appropriate for the Wilmington line or a search for other sources of technical knowledge. (Polidura did have a license from a German paint company supplying its technical know-how, but the company was acquired by Hoechst, which now has its own paint affiliate in Brazil, and the license has since expired.) The first move is to see what is available from Wilmington and what is necessary to adapt it to conditions in Brazil. So far, there has not been a 100 percent adoption of *any* paint by Polidura, simply because of raw material differences in Brazil. Also, Polidura's manufacturing lines having already been set up with a different type of equipment and a different range of products forces an adaptation of the U.S. lines. (If the company were starting from scratch, it would set up a quite different manufacturing plant than exists presently.) The third factor that forces an adaptation is the difference in customer needs.

From the standpoint of manufacturing, however, it would be highly desirable to adopt the paint line fully, since the paints have been fully tested in production and sales and would likely appeal to Brazilian customers. A prime part of the assistance from Wilmington, therefore, is to provide a full line of paints from which to choose. A further advantage of the association is assistance in the use of computer technology in paint modelling, manufacturing, and administrative services. A third contribution is in the managerial and organizational capabilities and assistance in learning what to ask of whom in Du Pont.

The primary role of the lab is thus in the substitution of local materials in the paint formulas supplied by Wilmington. For example, supported by Du Pont's technology in metallic paints, Polidura is seeking to shift automotive customers in Brazil to acrylics and away from the color/clear-coat system used in Europe. Some of them use the European paints based on enamel (both the color and the varnish are alkyd-melamine based) but this type does not resist weather well, especially direct sunlight. Polidura thinks that it is close to developing a final product satisfactory to its customers in Brazil, based on Du Pont formulations.

Much of Brazilian refinishing is done with nitro-cellulose paints. These are cheaper in Brazil than acrylics; though not as durable, they are durable enough for local customers. There have been no new paint developments in Brazil, and presently there seems little likelihood of the market demanding them. Therefore, there is no justification for much effort in applied research; it is simply easier to obtain any new developments from abroad or from a parent company. In any case, new paint products for Brazil probably would require imported materials. Except for water-based paints, paints are 50 percent petroleum derivatives (solvents) and 50 percent solids (to form the film). The Brazilian government is pushing the use of alcohol (derived from sugar and manioc) as a possible solvent. Availability will depend on the willingness of the sugar refiners and the national petroleum company to make these materials.

Another area of assistance from Wilmington is the process of decisionmaking. Procedures are not yet set up the way they should be, but new methods are being introduced concerning administration, manufacturing, marketing, and development, and an increasingly close coordination is being developed between the lab activities and marketing interests. The drive to adopt Du Pont management procedures is not only to increase efficiency and improve decisionmaking, but also to be able to communicate effectively with the parent organization.

Polidura has set up management-by-objectives with a month-by-month scheduling of the work program for each professional in the lab. Daily meetings of section heads in the labs are held to resolve operation and program coordination problems. Each of the project managers and the section heads have direct contacts with personnel at their level in other departments (marketing and manufacturing, for example). Twice-weekly sessions are held in which each group airs its specific problems with other groups to coordinate work on administrative and technical problems. In addition, there are separate sessions on safety throughout the lab. To help concentrate all laboratory work, technical services in wood finishes and refinishing are being shifted into the lab from the marketing area.

To prepare professionals and technicians entering the lab for the procedures that will be followed, formal courses on analysis and procedures are provided over a three-month period, covering materials, intermediates, administration, and organization. Another three months of on-the-job training is required before a new person can become useful.

Specific techniques that have been acquired from Du Pont include color technology—the standardization of colors and preparation for computer sharing—and pigment analysis, which involves the sending of color panels to the United States to be analyzed and broken down so that the Brazilian lab can match by hand. In addition, a considerable amount of safety information is provided, which is of great importance to the lab in reducing the hazards of handling raw materials and of the polymer processes, and in showing the need to use protective devices in various aspects of lab work.

In providing this assistance, Du Pont has provided three channels of information flow: the technical manager of F&F international operations, a contact point in the International Department, and a flow of technical reports and formulas, which reduce trial and error in the Polidura lab.

In addition to assistance from the United States, Polidura exchanges technology with Du Pont's Belgian lab. For example, Polidura has an air-dry refinishing product that needed to be upgraded to a heat-dry process (80°C). The Belgian lab sent a technician to explain the process, with the result that Brazilian professionals are copying it, thereby saving considerable development effort. A new line of refinishes would rely on the Belgian line, since it will be similar to European paints; U.S. types will be pushed later. The problem in this shift is that materials for the monomers used in acrylics are not produced in Brazil, whereas those for the alkyd enamels are available.

As to reverse flows of technology, Polidura is not yet helpful in any direct way to U.S. developments or to other markets. Though its work in polyester resins might have been useful to Du Pont, it was not at a sufficiently high level.

The primary need in coordination between the U.S. and Brazilian labs remains that of setting up appropriate operating parameters in Polidura to raise paint qualities, improve efficiency, and cut costs. The lab professionals have been highly receptive to these initiatives from Du Pont, and are learning quickly where and how to control quality. (For example, procedures will cut the cycle time in the ball mill by two-thirds, passes in the sand mill will be cut from five to two, and the sand level will be adjusted—all improving quality considerably as well as cutting cost.)

The Brazilian market is becoming more quality conscious, especially as the economy slows down and demand becomes more precise. Quality has not previously been a significant concern to Brazilian customers, who until recently had not demanded both durability and luster in their paints. For example, car paints look dull within one year, and the owners have simply accepted that fact. Even refinishers have not been interested in a higher quality paint, but only in the low price that attracts customers. However, this is beginning to change as European competitors arrive with better quality paints. Du Pont pushed quickly into quality lines, and the sales practice of the marketing division is beginning to reflect it; however, Marketing is still working hard, upgrading its efforts through developing its own sales force in place of a former agent system.

The government itself has not assisted in developing paint

standards. In fact, there are as yet no national standards for paint. Therefore, Polidura officials have not had to meet with governmental officials on regulatory issues.

Future of the Lab

Considerable changes in lab's projects and methodologies are anticipated as it accelerates the adoption of U.S. methods and line and coordinates more fully with Belgium. For example, the materials section of the lab now devotes 60 percent to quality control work, but it should expand fairly soon to improvement and substitution of local materials, without a significant expansion of personnel. The quality control work will probably leave the lab to become the responsibility of the manufacturing plant. The shifts will take place in order to be able to utilize Du Pont technology more efficiently than at present.

Among the project changes that will occur will be a new emphasis on water-based paints. There is no pressure yet from the government concerning pollution from oil solvents, though there is some concern over possible fire hazards.

The lab's position in wood finishes will be maintained, as new product development continues. And since formulations in refinishes are received virtually complete from Belgium, the only development work needed is in the adaptation of local materials. Almost all of this work is done by the marketing support group of the lab rather than a development group. Only one of the paints requires adaptation, and this is worked out in the development group.

One of the major reorientations yet to take place is in the relationship between the lab and the marketing divisions of Polidura. Lab projects have been determined haphazardly in the past; "product managers" will be assigned in the future to bring about the necessary coordination. The labs will be turned toward more developmental work related to the specific profit centers established within the company and to technical services supporting each product line according to cost/profit calculations. This should bring about a better allocation of resources within the lab, shifting the priorities among projects along with market shifts. The growth of the lab will then depend on the rate at which the Brazilian market becomes more sophisticated, along the lines in Europe and in the United States. The professional and potential technical competence of lab personnel is adequate to meet these needs, given U.S. technical support.

Another modification will be the injection of careful cost calculations in the lab work itself. There will be a manpower budget for the lab as well as costing of equipment, inputs, and evaluation of the outputs. Previously there was no lab budget at all, and no calculation of "development costs."

As a result of the changes made in the lab programs, the professionals and technicians have much greater room for development of their interests, allowing them to grow into more satisfying careers. Previously, when the lab wanted a particular technique, they simply went out and hired somebody with that capability; when it was no longer needed, the individual frequently left.

Some new product lines are likely to come out of the work of the labs—coverings for containers, for example, liners and exteriors for tin cans and wire, and other F&F items that might fit into Polidura's markets. Top management sees no sense in duplicating Du Pont developmental efforts, so the Brazilian labs will be predominantly oriented to reducing the cost of production, satisfying consumer needs, and adapting local materials, which have to be substituted for imports.

Du Pont's future growth in Brazil will, therefore, be in response to market development and shifts, and the growth of the lab will likely track with the process followed by the lab in Belgium.

Unilever Research and Development: A Study of International R&D Coordination

INTRODUCTION

Unilever makes a variety of product lines requiring R&D support: animal feeds, chemicals, detergents, edible fats and dairy products, frozen products, toilet preparations, meat products, sundry food and drinks, paper, plastics, and packaging. Coordination groups, headed by directors of the Unilever Board, watch over these industrial lines within Europe, crossing over national boundaries. They are responsible for a continuing flow of information concerning the products, market success, new products and processes, etc. The management of the national units in Europe reports directly to two regional directors, and a European liaison committee links the two regions. An Overseas Committee is responsible for all other countries, save Canada and the United States, which report directly to the top-level Special Committee of the Unilever Board. Unilever operating units, allowing for differences among countries, are, as far as possible, free to carry out their operations with a minimum of supervision and control.

The Research Division reports directly to the three-man Special Committee; the head of the Division is also a member of the twenty-two-man Board, along with two officials from the Overseas Committee, the two European Regional Directors, the nine Coordinators, and five other officials of the company.

Coordination of R&D activities has taken a long time to achieve in Unilever, since historically the company has had a "marketing" rather than a "high-technology" orientation. Responsibility for new product innovation falls to a large extent on the Research Division. But final responsibility for adoption and marketing rests with the chairmen of the operating companies.

Each operating company in the Unilever family sees itself as independent and self-contained. Loyalties of officials in Unilever subsidiaries are probably stronger to the affiliated companies than to Unilever itself. Unilever is a confederation in which the sovereignties of the affiliated companies are shared with the center, but only when it is helpful to the operations of the affiliates to do so. Consequently, centralization occurs when affiliates' problems become so similar that it is simply more economical and desirable to coordinate certain activities.

A Research Planning Group (RPG) is centered in each of the Coordinations. It recommends the budgets needed and the responsibility of each lab for carrying out research programs of interest to the Coordination. Each RPG is chaired by the R&D representative of the Coordination. The other members of the RPGs are the heads of the relevant central research labs or their representatives. A Science and Technology Strategy Group meets on an ad hoc basis for overall coordination among the labs; it is chaired by a member of the Research Division Executive Committee, with membership from senior members of the central labs.

The RPGs make certain that research activities support the business objectives of the company, attuning them to product needs rather than to mere marketing support. It is the responsibility of the RPGs to feed the business plans and objectives of Coordinations into research labs and to feed back into the business units the results of research projects. The overall business strategy for each product line comes from the Coordinators, with substrategies developed for product segments—for example, detergents for fabrics, hard surfaces, personal washing, etc. A Product Area Manager based in Research is assigned responsibility for each of these substrategies, in collaboration with marketing officials and others in each Coordination. It is his responsibility to obtain agreement with the operating units on the priorities for the brands to be developed. The agreed business strategy is modified by the collateral development of the "research strategy" developed within the RPG and vice versa, in the sense that the research strategy is modified by the business strategy.

Unilever research is international and coordinated from London. There are six European laboratories, which are run as six distinct administrative units (see Table B1.1).

The European labs report to the Central Research Division, which also maintains operating contacts with the two "affiliated" laboratories in the United States and the one in India, as well as some smaller units elsewhere.

Most of the operating companies maintain separate development labs, some of which are quite substantial. There are twenty-nine development labs in twenty-five countries outside of Europe, plus many more on that continent. The total number of people employed in all types of R&D labs is over 7000, including both professional and support staff. Of these, approximately 4000 work in the six major European laboratories, ranging from approximately 1400 at Colworth to 100 at St. Denis.

This study of Unilever R&D focuses on a few of these labs, visited over a period of some months: Port Sunlight and Vlaardingen, among the central labs all three affiliated labs, and the development lab in Brazil. In addition, interviews were conducted with officials of Research Division and Colworth, and with some R&D members of Coordinations and of the Overseas Committee.

RESEARCH STRUCTURE

The research division has three major responsibilities—administration of the research system, control over the research

Table B1.1. Unilever's European and Affiliated Laboratories

European	Vlaardingen/Duiven[a]	(Netherlands)
	Bahrenfeld	(Germany)
	St. Denis	(France)
	Port Sunlight	(United Kingdom)
	Colworth[a]	(United Kingdom)
	Isleworth	(United Kingdom)
Affiliated	Edgewater, N.J. (Lever Brothers Co., U.S.A.)	
	Englewood Cliffs, N.J. (Thomas Lipton, U.S.A.)	
	Andheri, Bombay (Hindustan Lever Ltd., India)	

[a]Heads of these operations are members of the Research Division Executive Committee.

programs, and coordination of the various scientific projects undertaken by the labs. Research programs such as that on skin cut across several operating companies and product lines, and require substantial coordination to prevent duplication and to increase their usefulness.

Organization

The Research Division Executive Committee includes the Head of Research and the heads of two European labs; it centralizes policy control over several multiproduct, multidisciplinary, multinational laboratories, with each laboratory able to make inputs to each product group according to its capabilities and interests. Research Division can use Central Research funds for specific projects undertaken by one of the labs. These Central Research projects will be added to the annual program of each lab, using additional resources at each.

The budget for the labs is agreed on with each of the Coordinators and divided by the head of the Research Division among the six labs. Each company pays a percentage levy on sales set for each product group. These percentages will differ among Coordinations but not among companies that form any one Coordination.

The line authority of the Research Division extends to Europe only, thereby excluding Overseas and the United States and Canada. The Research Division could extend its control of policy to overseas (but not U.S.) laboratories, but it has decided not to do so, partly because it is difficult to get the attention of, or pay attention to, labs that are so far away. Consequently, the overseas labs remain tied closely to the local operating companies, and are categorized either as development labs or affiliated research labs. The latter have evolved from development labs and have tied some of their program to that of the Research Division and receive some support from it.

Central Research funds, mentioned above, amount to about 9 percent of the total budget of the Research Division labs. The rest of the funds come from the Coordinations, plus small amounts from other specially interested parts of Unilever.

Structure of Labs

Basically, two types of labs exist: Central Research labs and development labs, with "affiliated" labs having characteristics of both.

Central Research Labs. The Central Research labs are organized to coordinate the inputs from various disciplines with the needs of various product areas to provide outputs that are useful to the various manufacturing companies. They are also organized to produce reports on scientific and technical developments in each discipline area of the lab that will be useful to the other labs in their work. Each lab is a complex unit involving multiple inputs, multiple outputs, multiple controllers, and multiple users. For this reason, systems of matrix management are highly developed.

The multiple *inputs* are from the various scientific disciplines such as organic chemistry, biology, microbiology, analytical chemistry, and mechanical engineering. For example, the Colworth Lab has some of the best biochemists, microbiologists, toxicologists, and animal biologists in the world. Its professional personnel are internationally renowned, and the output on food research is greater than the combined research efforts of four other European countries. Consequently, the Colworth and Vlaardingen labs are together recorded as a single country in EEC statistical tables on food research. The capacity in the Colworth lab is greater than the rest of U.K. capacity in this field, permitting it to address *any* problem in foods.

Some of the inputs into each lab begin outside in the marketing divisions and others inside the labs from analysis of potential products by professionals. A research project proposal is written for each idea generated, and the Research Planning Group (RPG) chooses the lab projects it considers suitable. It then divides them according to the interests of the lab, assigning to some the project responsibility, to others segments of the projects, and shifting the responsibilities through the different phases of the project as appropriate. The labs assign project teams of two or three professionals in one or more labs, and if coordination is necessary establish procedures for monthly meetings.[1] The program/project matrix for a laboratory is shown in Table B1.2.

The research *output* of Unilever falls into three main branches of applied science—foods, detergents, and toilet preparations—with each of the major projects falling into these three categories:

Food Science. Edible oils and fats, frozen foods; dehydrated foods; canned foods; dairy products; cooked meats, bacon, and sausages; fish products, nonalcoholic drinks; animal feeds.

Detergent Science. Detergents of all kinds, including domestic and toilet soaps; washing powders and rinse products for fabric; dishwashing products; hard surface cleaners; detergents for industrial use.

Table B1.2. The Programme/Project Grid

Research Project Managers

Research Resource (grouped into major section programmes)

Research Projects (grouped into categories)	Dental Research			Hair Research			Skin Research	General Research Division		£
	Product Formltn.	B'ground Techgy.	Evaluation	Product Formltn. I	Product Formltn. II	Evaluation	Prod. Form. & B'ground	Explor. Science	Techgy.	
Dental Products										
Product Projects	●	●							●	X₁
Background Formulation		●							●	X₂
Dental Research			●					●		X₃
Technical Support			●						●	X₄
Hair Research										
Product Projects						●				X₅
Evaluation Research				●	●	●			●	X₆
Hair Research						●		●		X₇
Technical Support						●			●	X₈
Skin Products										
Product Projects							●	●		X₉
Skin Research							●			X₁₀
Technical Support							●	●	●	X₁₁
Quality Standards			●			●	●		●	X₁₂

Product Area Managers

204

Table B1.2. *continued*

Research Project Managers

Research Projects (grouped into categories)	Research Resource (grouped into major section programmes)									£
	Dental Research			Hair Research			Skin Research	General Research Division		
	Product Formltn.	B'ground Techgy.	Evaluation	Product Formltn.I	Product Formltn.II	Evaluation	Prod. Form. & B'ground	Explor. Science	Techgy.	
Laboratory Initiated & Corporate										
Materials Research								●		X_{13}
Biophysics								●		X_{14}
Physical Chemistry								●		X_{15}
Skin Biochemistry							●			X_{16}
Inter Lab Support								●	●	X_{17}
Planned Expenditure £	Y_1	Y_2	Y_3	Y_4	Y_5	Y_6	Y_7	Y_8	Y_9	Z

Product Area Managers

Legend: A simple matrix (adapted from Reference 3)

● annual cost of section resource committed to Project

X annual cost of total resource committed to Project (X_1 not necessarily $\approx X_2$ etc.)

Y annual cost of section resource ($Y_1 \approx Y_2 \approx Y_n$ for convenience)

Z annual cost of laboratory

Note that $\Sigma X = \Sigma Y = Z$

A Laboratory Matrix

Toilet Preparation Science. Hair, skin, and dental preparations, including shampoos, conditioners, hair sprays, hair colorings, toothpaste, and other toiletries.

These multiple research results are recorded in over 2500 assorted internal reports each year—some very specific to a single product, some with wide applications in business, and some making substantial contributions to the advancement of science and technology in general.

The reports are prepared by the professional scientists on the part they play on each project—the vertical axis of the matrix; by the Research Project Manager on the progress of a given project—the longitudinal axis; by the Product Area Manager on a periodic basis to the RPG (vertical axis); and by the heads of the lab to the Research Division. Besides the obvious uses of these reports, the final objective is to provide improvements in processes and products as well as new products to the manufacturing companies, which are principally in Europe, but also throughout the world. Each Product Area Manager is responsible for liaison with the operating companies on an individual basis, explaining what is being done in the research and how the company might follow up.

Each lab is organized with Division Managers who are responsible for several disciplinary sections, such as chemistry and analytical chemistry or biology and microbiology, with a Section Manager responsible for each of these disciplines. The Research Project Manager is responsible for the activities of *all* of the disciplines working on a particular project; he may hold this responsibility while also being a Section Manager. Similarly, a Division Manager may also be assigned as a Product Area Manager, looking over several different projects as they fit into product needs.

Development labs. The division of labs into those within the Research Division and those tied to the operating companies (development labs) reflects a policy decision to pursue two different research routes simultaneously: one that concentrates R&D on one product line in a central location serving all manufacturing companies related to that line, and another that develops expertise on several products serving the national markets.

The company development labs are independent of the Research Division and the RPG's in their work. Only the Edible Fats and Dairy (EF&D) Coordination has achieved substantial control over company labs dealing with its products, reflecting a desire on the part of the parent company to shift control from regional administration to Coordinators. The latter are strengthening their hold on product development but do not presently seek "divisionalized" research. The EF&D Coordination is seeking control over brand introduction in the various markets, so it wants control over the development of the products, hoping to make them similar and dovetail their promotion. The Detergents Coordination is following the same route in some areas.

This pull toward centralization is matched by one towards decentralization. The Australian lab broke up its facilities for basic R&D to return to the status of a company development lab. And the Lever Brothers lab in the United States has separated the research from the development function, due to a reorganization of the company into business units; the development work has moved towards increased productline organization with direct responsibility to the manufacturing profit centers.

The independence of some of the development labs overseas reflects a desire to have them locally oriented, serving local markets and employing local scientific and technical personnel. In addition, the overseas labs are technically better oriented to local product needs and market demands, which would be obscured to a central research lab. This point is more critical for a consumer-oriented company than for one that might serve industrial customers. The product development labs, therefore, need to be close to the consumer, in Unilever's view. However, basic research is preferably centralized at a few locations; basic research facilities are not needed everywhere, and company officials do not think they could successfully coordinate research dispersed over more than five main centers.

The company development labs have over 100 employees at the largest, and 10 to 15 at the smallest. The responsibilities of these labs are to provide developments for new products and the optimum use of new processes, and technical services and quality control; budget and manpower are usually split 50-50 between these. The total budgets for these labs are determined individually by the cost center to which they are attached. There are over 100 development labs, plus over 400 companies that have only technical services and quality control facilities.

Every plant requires some scientific and technical personnel in technical services and quality control activities. Establishment of a development lab occurs with the expansion of production and a desire on the part of the affiliate to express some company pride, based on a presumed differentiation of markets or processes, plus pressures for the use of local materials. The total budget for the company development labs amounts to about 30 percent over and above the budget for the central labs.

"Affiliated" labs. Development labs attached to the various companies are eligible to gain "affiliated" status, which provides them with some of the advantages of Central Research affiliation without pulling them from the control of the manufacturing company. (To move from development lab to affiliation requires a decision by the head of Research, the Special Committee, and the Overseas Committee.) A development lab is ready for affiliated status when: (1) it becomes politically desirable to improve the status and image of the lab (as is expected more and more in the future); (2) the lab is technically sophisticated or must support sophisticated products, and has developed an adequate base within the lab; (3) the market has become sophisticated enough for different products; (4) government regulations require local testing and evaluation; (5) the lab is highly specialized and of a critical size to support more than developmental research—no less than six specialized professionals with, say, twenty support personnel; (6) when its program merits some central funding.

The U.S. labs and the one in Bombay, India have reached this status. Those in Turkey and Indonesia are under consideration, though no initiative has come from them. Rather, the heads of the Research Division and the Overseas Committee began to consider them out of a concern for the need to show their interest in development of science and technology in developing countries. Normally, the initiative would be taken by the local manager, who should perceive the need for a higher status, though he may not, not realizing the benefits of having an R&D lab.

In practice, it is not difficult to shift from development to affiliated status, the biggest problem being decisionmaking within the operating company itself, which may see a loss of some control. The head of the development lab actually gives up nothing to achieve affiliated status, since he must continue the development program. He simply gets more from the Research Division and is under its surveillance for scientific quality control, which may produce some headaches in his research management, but which his program needs to undergo.

The local manager of the operating company loses some control over the development lab, whose program would now be reviewed by the Research Division. Although this is ideally the case, it does not represent the situation in India or the United States, where the Research Division does not exercise this type of control.

In fact, it is not to the advantage of Unilever to have the local manager lose sovereignty, for he should be able to claim publicly (and correctly) that "My lab achieved. . . ," despite any background assistance from the Research Division. Since the Research Division is not itself a profit center, but a cost center and a "funded service group" in the company, it does not get credit for increasing the company income; rather, this is credited to the manager of the operating plants.

The objectives of the affiliated lab remain quite similar to what they were previously, but its research program shifts somewhat from empirical efforts to meet specific problems to more deductive analysis relying on research efforts elsewhere. The scope of the programs also changes somewhat because of closer ties to the local university and science community.

Contributions from the Research Division arise in the exchange of research reports, technical visits of personnel, and assistance on quality standards, as discussed below. In addition, the chairman of the appropriate RPG would visit the overseas lab, but not on as formal a basis as if the lab were within the central research system.

The cost of affiliated status is borne by the local company, but as a part of the central levy which is paid to the parent company annually. Thus, affiliated status increases the contributions coming from the Research Division without raising costs or affecting the profit picture of the affiliate. Though costs seem to be borne by the local company in its payments to the parent, they are matched by a reduction in returns to the parent company. Therefore, the costs of affiliation are essentially nil to the local manager.

The experience in India (discussed later in this case) indicates some of the benefits of affiliation. What occurred there is a greater adoption of products based on developments in the Research Division, with the effect that the central research is now more closely supporting local production objectives. This is a happy marriage of mutual objectives.

Operating managers think they can tell R&D labs what to do and frequently do so. Instead, they must find out what the labs can do in relation to what the company thinks needs to be done. The R&D lab cannot simply seek to serve marketing needs as determined by the marketing divisions. Rather, it has to search out the actual needs of the consumer and find out how these can be met. The result may be a change in the product line rather than larger sales of an existing line. This shift may appear to threaten the manufacturing company's manager, for it forces him to ask questions about consumer needs rather than merely how to produce efficiently and sell existing products.

The shift from marketing and market research to consumer research and R&D is a difficult one. Market research relates to pricing and quality perception on the part of the consumer. Consumer research relates to specific consumer needs, uses of the product, and changes in broad demand among consumers. In some countries and in different industry lines, consumer research is carried out by the R&D labs, while in others it is done also by the Market Research Division of the operating company. The different approaches depend on perceptions of roles and specific capabilities.

One of the problems in making affiliated status effective appears to be that of distance, which makes communication difficult. But, according to Unilever officials, geographic distance is not the greatest of differences, and does not produce the greatest difficulties; rather it is cultural differences, which can be closed only by frequent visits both ways, as will be seen in the Indian case. However, the geographic and time distance does reduce the number of visits to India as compared to the United States or even South Africa, where the time change is not significant. Similarly, lab officials observed that visits to Japan and Brazil are inconvenient and are less frequent than they probably should be.

The final shift from affiliated status to that of a central research lab is not worth attempting until the scientists in the lab become sensitive to what is involved in basic research and feel the need for affiliation with a larger scientific community. This step seems a long time away for Unilever overseas labs, even those that are presently affiliated. Differences in consumer goods tastes and product lines make centralizing consumer research inefficient. The labs are too distant and too expensive to visit for long enough to achieve effective coordination, and an effort to centralize would not be made until they became too sophisticated to manage locally. It is difficult to achieve the concept of "living in the same ambience" with others in the scientific community across long distances and substantial cultural differences.

Advantages of Affiliation with Central Research

Within Unilever, communication ties, the support relationships, coordination, and information exchange are quite different according to whether the labs are associated with Central Research or are development labs. The facilities of one Central Research lab are available to another for the mere asking. However, since resources are already committed, the RPG may have to be called upon to persuade one lab to reallocate resources to help another. Assistance may be asked by one lab of any of the others on any given task, drawing on specific expertise. In order to achieve this exchange of assistance, an intricate network of communication has been established, which is so good that it is used also by nonresearch personnel. The advantages of being a Central Research lab include:

Communication of research results and success of projects
Management development and training of technicians
Seminars on current issues and research results
Supplementary funding for Central Research projects
Expert assistance in sorting out good ideas from bad, and group participation in project design

The most important of the different types of communication is the circulation of research reports on each project, which are available to all scientists in each of the Central Research labs. This provides multiple inputs to any given problem and creates a "unity of belonging" to the company's science community. Secondly, personal visits occur on a frequent basis (for example, about seventy per year between Europe and the United States), though they require approval of senior officials. Third is communication into the lab from various business units concerning their product needs, to ensure that lab's work is more effective in meeting the demands in the various product areas. Fourth, direct communication exists to all labs, facilitating telephone conversation. Unilever tries to make telephoning easy, leaving the decision as to calling or writing with the researchers.

A publication entitled "Research News," which summarizes project reports, is sent automatically to all labs. Many reports are also sent automatically to labs that have expressed an interest in particular projects. "Research News" is designed to stimulate scientists to ask for reports in which they are interested. Since widespread communication is encouraged, professionals are often in such current contact with each other that they are *more* up-to-date than the reports. The reports then become more a matter of record than a stimulus to inquiry. Sometimes, however, they cause a researcher in one area to offer help to another in an unrelated area.

In addition, a Technical Information Service (TIS) is run by the Management Services Section of the Engineering Division. This section channels a variety of reports, including those from the Research Division, and screens them for distribution to the various operating companies. This type of communication helps the companies know who among the Central Research labs would be useful in case they have a particular problem.

It is not always easy to find evidence that the channels of communication are effective in producing more desirable results. There is ample evidence that communication occurs, but to make a cost/benefit calculation is much more difficult. There is also considerable evidence that the research professionals know of the work of others in similar fields and that they are in frequent telephone communication. That there is very little duplication of research as compared to ten years ago is further evidence of communication, since it could only be stopped by personal visits at several levels; otherwise, each lab would simply proceed along the lines that it considered to be of its major interest.

Removal of this duplication is the responsibility of the head of each lab, who is increasingly willing to rely on another lab because he knows the quality of the work there and the expertise of the personnel. This personal knowledge is achieved by week- and month-long visits that permit the development of mutual understanding. Shorter visits of one or two days can be useful in skimming off developments in another lab to permit continuing exchange of information later. The ability to eliminate duplication by top-level decisions is feasible only because of close relationships at lower levels in the lab, so that professionals will in fact rely on the work of others. For example, the lab in India had an idea for a new project on skin research; misunderstandings with central labs resulting from written communication over two years were sorted out only by two personal visits of top scientists. Detergent research was begun in India through personal visits from Port Sunlight, and that on product safety and toxicology was begun from similar visits by Colworth scientists.

The second advantage of affiliation is in management training and exchange, which is so normal among the Central Research labs that records are not kept in a fashion revealing their frequency or content. As employees' lab responsibilities increase, their abilities are checked to see if they can profit from foreign experience. If so, they are loaned to another lab either inside or outside the Central Research system. The R&D managers in Brazil have been loaned for over six years at a time, and men have been loaned to India at the assistant manager level and to Turkey.

The third advantage is the opportunity to participate in seminars on recent developments. Each lab has a program under which experts from other labs or outside the system entirely are requested to present papers on their research, sometimes on highly specialized subjects such as infrared spectroscopy. On occasion, when it is feasible, officials from other labs will attend these seminars. Since some of the Central Research labs are quite large and have diverse expertise among their professionals, seminars can be given by scientists within the lab, providing insights into developments in various disciplines.

The fourth advantage is the addition of some 10 percent to the budgets of the Central Research labs. These Central Research funds are meant to sponsor projects that do not fall automatically within the interest of specific Coordinations. Unilever officials expect that this activity will increase as they learn more about how to achieve the results they are seeking. Few of these funds have been extended to labs outside Central Research—India was allocated Central funds of $50,000 on skin research because it would have application elsewhere.

The final advantage of affiliation—group design of projects—results from the decisionmaking and control structure described above, which marries research capabilities and business opportunities. Communication between the sections of the company having responsibility for business opportunities and R&D labs is absolutely necessary. There must be a translation of R&D results into nonscientific language, which can be done only by top-level (and highly capable) professionals. These professionals are so competent in their own fields and confident of their own understanding of the science that they can "play around with the story" and tell it in different ways to top management without doing damage to its accuracy.

Shifts in Structure

The lab of the Canadian company illustrates the way in which structure and programs must be adapted to changing situations. The lab was first set up in the 1950s, despite the company's existence since before the turn of the century. It soon found that its projects (and products) were quite similar to those in the United States, and that no product development was needed different from what was already being undertaken in the United States and Britain. It was then shifted to guidance of the Research Division under pressure to do some basic research,[2] and the incentives in tax relief offered by the Canadian Government were used to justify the effort to make some independent contributions. The lab embarked on a research program in skin, dental products, and flavoring, using Canadian scientists. The Canadian government dropped its support after five years, and Unilever had to collapse the lab, since there was no real use for the results it was producing. The lesson it learned from this experience was that an R&D lab cannot be created and sustained without a product rationale, which the Canadian operation did not have. There was a belief that half-price research was bound to be useful, since "all R&D is useful sometime, somehow." The company found that it is not, and that research should be located where there is a possibility of product application, to emphasize the usefulness of the research and to give it urgency.

The results of basic research have to be pulled into production. Such research is done by a group of specialists who can generate background data and results which may become useful in many different areas. Getting the ideas from basic research into gestation of products for further development is extremely difficult, however. There are a thousand ways *not* to do it and only one way in which it *may* be done. What is needed is a strong pull to draw out the results of pure industrial research into the commercial stream. Without urgency, the research may never be completed. (Academic research, on the other hand, is completed for the mere purpose of publication.)

This problem of the flow of R&D from conception to utilization raises a question of organization: centralization versus decentralization. The decision to favor one or the other is frequently not based on rational factors. The cutoffs in lab sizes and program content are related to the factor of critical mass for specialized activities (for example, spectroscopy) and limitations on managerial control at the top. In between, the decision relates to market size and the complexity of products or processes and the size of the resource base. From Unilever's experience, a minimum of 300 staff members are needed for broad-scope laboratory research, while it appears that the number can reach 1500 before a lab gets out of control. One of the critical factors is the cost of information and its dissemination and control. Information needed by a very specialized lab may be exceedingly costly and difficult to sieve. On the other hand, information flowing into a complex, broad-scope lab is also difficult to channel to those who need it and to be certain that they use it effectively.

An Australian research unit existed from 1954 to 1962 that was considerably more than a development lab, though it was not affiliated with the Research Division. It was broken up because the Overseas Committee wanted profit responsibility located strictly in the individual Australian companies, including control over their R&D activities. Consequently, 150 employees were shifted to the labs within the operating companies, redirecting their focus to product development. A very small specialized unit was left (six professionals in a total of eleven employees) to provide central research services, with close links to Research Division.

CENTRAL RESEARCH LABS

The programs of the central labs are composed of pure and applied research; developmental research, as necessary to get the operating companies to understand the uses of the research and pick it up; engineering, including a pilot plant and test materials; and finally, technical services. These labs are the source of most new products or processes, and they have the responsibility to demonstrate production feasibility. For example, a lab might develop an idea of frying foods without fats; once the process was shown to be scientifically and technically feasible, it would develop prototype machinery, and the engineering division might become involved in production of the machinery and installation.

The technical services supplied by the central labs involve research on "safety-in-use" on all products. The labs also provide analytical services on breaking down competitive products to determine their composition and whether there is an infringement on Unilever patents. Services are also extended on problems such as corrosion of pipes and equipment, the analyses of environmental pollution (measurement methods and testing), and the testing of workers' health.

The breakdown of the program of the European labs under the Research Division is more or less as follows:

Background and basic research	25%
Product development	30%
Process development	15%
Safety research	10%
Consumer research and product evaluation	10%
Technical services	10%

The programs in pure research require a critical mass of specialized professionals, centralized in specific labs to reinforce each other's work. For example, pure research in edible fats requires the following inputs: agriculture research on sunflowers and rape seeds, biological research, background research on fat processing, physical chemistry, organic chemistry, research on milling and crude oil handling, automation and control of equipment, product development, and packaging. All of these must be coordinated by the Research Division Executive Committee.

To dissociate basic research from the other steps in the process down to product development and commercialization would be to remove its urgency and give the wrong signals to the scientists. They must feel part of a process that requires results in a particular sequence and time frame. Consequently, Unilever officials have stressed the importance of the link to consumer research, constructing a chain back to the research design. Still, there is a temptation to seek solutions prior to problem identification. Problem identification will be especially necessary in foods, where no one has yet answered satisfactorily the basic questions about man's nutritional needs. Emphasis on problem posing has occurred in detergents and toilet preparations as the result of the pressing competition from Procter & Gamble and others in Europe, who challenged Unilever companies in their seemingly protected market; even so, some operating companies would not move before they got this competitive signal, despite prodding from corporate officials.

Finally, the Research Division attempts to supply what is lacking in many research organizations—the marriage of a scientific mind with management capabilities. Labs can get all of the scientists and technicians they want to solve given problems, but the supply of problem identifiers and good management coordinators, who can see that projects get done, is scarce. The answer is not to put pure scientists in managerial positions, but rather to put managers with knowledge of science in charge of R&D labs; such persons seem to be hard to find. Therefore, husbanding of scarce managerial resources is necessary and can be done through centralization.

Only two of the six central labs—Port Sunlight and Vlaardingen— were visited in this research, but there was an opportunity to talk to officials from Colworth, Isleworth, and Research Division, as well as some of the R&D Coordinators.

Port Sunlight

The lab at Port Sunlight is on the site of the original soap factory, and was attached to the manufacturing company at an early date, though it was some years before the facility became a separate R&D lab. It presently has a total staff of 920 persons, divided roughly into four equal parts of scientists, assistant scientists, technicians, and administrative and support personnel.[3] The scientists are the prime movers, with about three-fourths having Ph.D.'s; the remainder have B.S. degrees but are qualified to undertake Ph.D. degrees. (Evidence of their professional capability is seen in the number of publications in professional journals—forty-six in 1974 and thirty-two during 1975.) The scientists are in charge of identifiable segments of the total lab program—several related projects or a single project. The assistant scientists are in charge of distinct parts of projects.

Program. The research program at Port Sunlight (PS) is mostly in detergents. Its main objective is to expand the commercial base for detergents throughout Unilever. The lab also has an engineering division for examination of processes for handling effluents.

The program on detergent research extends from basic chemistry to consumer research (how the housewife uses detergents), with a constant eye for gaps in the product line to meet consumer needs. (PS's consumer research contains a continuing collaboration with the developmental labs and the marketing divisions in the operating companies not only in the United Kingdom but also in Germany and France.) Washing problems involve some highly sophisticated chemistry, and the analysis can be done appropriately for the consumer only after careful investigation of how the consumer washes and what is washed.

The researchers also look for problem areas such as bleeding of colors, bleaches, and various cleaning uses of detergents in order to create research programs which are "needs-driven." There is little "blue-skying" in the research on detergents because the professionals feel that they know the needs and problems in that area better than anyone else; new ideas from outside would be rejected as unlikely to meet existing or prospective needs.

The development of research programs proceeds as follows: First, a research theme is generated, creating a Research Project Record (RPR) from, for instance, a recommendation by the RPG for a "skin benefit bar" to both clean and improve the skin. The lab then constructs the methodology for pursuing the research and the timing for the project. The theme—skin benefit—requires an analysis of skin itself and how it may be affected, what chemicals may be used, and what biological effects are likely; the theme continues on to the precise product means of supplying the beneficial chemicals. Care has to be taken that no single part of the theme becomes a dominant project. For example, the single project on creation of the bar itself could be weighted so heavily in the research effort as to become dissociated from the background research concerning its actual benefit to the skin. Not only must the total research theme be coordinated, but so must pieces within it, as in the coordination of all background research to examine carefully the basic problems of surface chemistry on the skin.

Such coordination of basic themes means that there is a limit to the number that can be effectively managed. At the time of the study interviews, PS and sixty major pieces of research, which the management would have preferred to be reduced to about twenty. (In comparison, the Colworth lab had 600 much smaller projects, reflecting a different philosophical approach to research based on the needs in sundry foods at the developmental level. The emphasis on development, in the eyes of some Unilever officials, could lead to missing some basic similarities behind food needs, which would be surfaced by greater emphasis on basic nutritional requirements in humans and animals.)

The program at Port Sunlight goes through product formulation, providing a product and its specifications to the company, though the lab does not necessarily completely check out the product in terms of mass production. The company receives a "kit" for production, including the report on the materials, processing, and projected effects on the skin. The lab does not provide the company with all of the backup research or alternative investigations it made prior to the development of the particular product, but it does provide enough such information to help the company protect itself on advertising claims and safety procedures. (Further safety and toxicity tests would be done at Colworth.)

The operating company receives the research package and will rerun some of the lab experiments and the pilot phase. (This reflects partly an NIH factor in the company's development lab and partly a safety factor in rechecking.) The company then moves to large-scale pilot production, still on a test basis, since the processes and materials differ when scaled up. (Things are different when combined in larger volumes; one batch of a product caught fire when put into large-scale production.) In addition, different plants have different equipment and processes built into the factory and cannot handle some of the formulations. Each formulation, therefore, has to be checked out in the existing facilities. Where modifications are needed in detergents to suit the equipment and local materials (which may not be of the same quality as used in the lab), they are frequently made at the plant. The final product, therefore, is not always identical to what was developed at the lab and may not meet the consumer needs as originally researched, meaning that some of the background and even safety tests used for purposes of obtaining approvals of the product by regulatory authorities do not remain valid. (Even a change in toothpaste flavoring will invalidate test data for regulatory purposes.)

If significant changes are required by the factory, the development lab may throw the project back to the central lab; otherwise it might make small changes, depending on local expertise and the difficulty of the changes.

The last major segment of the PS lab is the engineering division, which has some 100 engineers specialized according to product group or function. All of these engineers work on design tasks, supporting the engineering functions in the companies. If a company needs new equipment or a new plant, the engineering division will offer advice on design and procurement in collaboration with the central engineering division of Unilever. The engineers sometimes specialize in quite specific problems, such as that of corrosion that effects the life of materials in the plant, or effluent treatment. They have worked on the science behind the problem of corrosion and wear in materials and have developed "reliability engineering specialists," who are especially helpful to companies during construction phases. Instrumentation control specialists also guide the companies on the selection of equipment. There is no set standard for equipment in producing any particular product line, and since Unilever's companies use many different kinds of equipment, an engineering report on one plant would not necessarily be useful at another.

Reliability engineering also requires a chemical input and examination of processing alternatives. These differ from country to country, since no single detergent processing strategy is applicable to all.

At one time there was an effort to provide a centralized service on engineering and construction to prevent empire building in the plants. However, the plants have, for a variety of reasons, insisted on their own architects and construction engineers, so the engineering division is left with highly specialized services.

Cooperation and Coordination. Historical specialization among the central labs has been continued for the most part—for example, machine dishwashing and biological expertise at Vlaardingen and fabrics washing at Port Sunlight. Coordination of project work among the major labs is often done by "horse trading," though in some cases an especially high expertise exists in one lab. In addition, some project units have been transferred from Port Sunlight to Vlaardingen to make room at the former for expansion in new directions. Where joint projects are undertaken, they are placed under a single coordinator in one of the major labs (usually a section manager or senior scientist). For example, the evaluation of detergents on fabrics requires the use of capabilities in each of the major labs, so different methodologies are developed at each and exchanged.[4]

Cooperation in projects with the Lever Bros. (U.S.) lab at Edgewater is infrequent. In the late 1960s, joint projects attempted in several areas—detergents, toiletries, etc.—generated some difficult communication problems. The results were not satisfactory, partly because of a lack of clear definition of the projects' objectives. In these cases, the U.K labs were doing work for the U.S. labs, rather than there being a true joint project, with clear objectives and a single coordinator. In a project on low- and non-phosphate detergent formulation and testing, the different washing machines and water temperatures used in the United States and Europe colored the research orientations of the two teams; too, the existence of phosphate-ban areas in the United States and not in Europe altered the research priorities of the project managers. Each coordinator was required to visit the other lab every three months to try to smooth things out. One product developed in this way finally got into test marketing, while another has not yet reached this stage; both are being developed in the United States. Officials at Port Sunlight believe that the major problem was poor definition of consumer need, resulting in a bad definition of the project itself.

A substantial effort at PS now supports the U.S. lab. Communication with Edgewater is affected by the U.S. lab's being attached directly to the operating company. (In comparison, PS-Vlaardingen communication is much better and more continuous, each being a part of the Central Research system and the research planning groups.) Even if there were more cooperative efforts, PS officials do not see that either lab would want to develop common products under joint research programs. Since PS and Edgewater have different research philosophies, as seen in different definitions of goals and methods and different product lines and different material inputs, it is difficult to set up joint research projects.[5]

Since India cannot import oils such as those used in Europe, it cannot join other labs in an extensive cooperative project. It can use the technology on refining, bleaching, processing, and evaluation but not on materials inputs. Therefore, PS can help with methods and the research management but not in the actual research. Of greatest importance is keeping a flow of information into India, maintaining continuous communication on specific activities. This is done best through personal contacts, yet even these will not produce a complete meeting of the minds between Englishman and Indian.

One example of successful transfer has been in consumer research techniques. PS sent one professional and a section manager to Bombay to help the Indians learn PS approaches and the significance of consumer research, and they feel that this was successfully picked up by the Indian lab. There have been few requests from India for significant assistance in the form of work to be conducted in Port Sunlight. The lab has sent out specific technicians and professionals, but considerable adaptations have to be made—even in the testing of efficient washing, for example, since washing techniques in the two countries are quite different.

Four individuals at the section manager level at PS have been appointed as counterparts of Indian lab officials (one has since visited India) in order to directly supply information outside of channels; private correspondence helps explain procedures more fully. But it remains difficult to coordinate work over such a distance and through the largely separatist mental attitude on the part of HLL managers.

To provide assistance to overseas Unilever companies ("overseas" is anything outside Europe or the United States), PS has set up an Overseas Section that currently has eight scientists, three assistants, and ten technicians. They have five projects, of which two are background, one seeking to accelerate innovation by transferring modern technology, a second assisting in the rationalization and optimization of brands to meet specific consumer habits in the various regions served. The other three are concerned with service to companies with technical problems, specifically Brazil and Japan.

The Brazilian project addresses the improvement/introduction of new products in the fabric washing market. Brazil has particular responsibility for consumer research and evaluation; Central Research supports this fieldwork with methodology and background scientific studies. The program is executed jointly under the control of an expatriate Development Manager (an ex-PS Research Manager) and the PS Overseas Section. Unilever contributes two Research Scientists to the joint project team; under central guidance, the team has set up a novel methodology for consumer research guidance – consumer testing and monitoring procedures in the field—and has developed innovative technology (in the overseas context) to solve anticipated and proven consumer problems. The results of this project may be utilized in a number of other overseas markets.

The joint project in Japan is similar in that the Japanese company supplies the research on consumer needs, while much of the technical work is done at Port Sunlight. Communication in this project is easy because there are two Britons in Japan, and because the project is not yet at a complex or difficult stage. They expect more difficulties when they reach the manufacturing stage.

Vlaardingen

This study of the Vlaardingen lab also focuses on its coordination with other labs, especially those outside Central Research. A brief description of its capabilities is necessary, however, to illustrate the types of specialization and integration that could occur.

Capacity and Programs. Vlaardingen ranks among the best of labs of all Dutch companies. Its status among scientists would be considered among the highest, and it attracts the best biochemists in the country. (Philips would get the physicists, and Shell the top physical chemists.) Vlaardingen must compete actively with other companies in Holland only in the process engineering field, where it is among the best, competing successfully with Shell for university graduates. On the international scene, Vlaardingen would compare with the laboratories of General Foods, Nestlé, and CPC.

The Vlaardingen lab is headed by a chairman, who has three deputies in charge of three conglomerated divisions: one includes nutrition, organic and biochemistry, mathematics, physical and chemical analysis, edible fat products, and catalyst and porous structures. A second includes process engineering and development, detergents, physical chemistry, the research and development application unit (RDAU) on detergents, personnel, and safety and security. The third includes flavors, proteins and microbiology; engineering and instrumentation; scientific information; accounts and establishment; and the Duiven site, which includes food science and agriculture and technology. The research program at Vlaardingen is concentrated in fats and oils, foods, and detergents.

The lab was begun in the 1950s to create an R&D presence on the Continent to meet different consumer uses and tastes, and to develop different approaches to research. (There was also a belief that the optimum size of a lab was between 500 and 700 persons, and beyond that labs should be divided. The lab at Duiven in Holland was spun off later for this reason.) Since Port Sunlight had begun research in margarines, oils, and fats, both the Dutch and British companies now had a base across all research fields. Since both labs are in detergents, coordination is very strong and centralized, using international brands, standards, and specifications to reduce costs and to meet competition. Cooperation in detergents also implies specialization, dependent partly on the expertise of scientists at each laboratory. Thus, Vlaardingen is stronger in biochemistry and PS in physical chemistry.

As in all central labs, coordination on individual projects is the responsibility of the Product Area Managers, who formulate the objectives of two or more main projects within their group and coordinate the division of responsibility between the two labs, keeping them reasonably disentangled from each other—a responsibility seen in the horizontal axis of the matrix in Table B1.2. The responsibilities of the PAMs extend across all project inputs, but they will not necessarily make contacts with the operating companies. The PAMs have different management styles, so that some are quite effective in making contacts with the companies and others are not, or may not find it necessary.

Coordination and Communication. Liaison with the U.S. labs is completely open in terms of the exchange of reports, meetings of the heads of labs, and cooperation on specific projects. For example, there is complementary work on the components of tea in an effort to isolate its peculiar flavor. The lab at Lipton has the right to participate in the deliberations of the RPG on a consultative basis, but it does not do so, since its work is not that closely related.

Only a few formal ties exist between Vlaardingen and the U.S. labs. Both are in vegetable protein research, particularly soybeans, in an effort to make it suitable for a variety of uses. The basic quality of the soybeans is not yet good enough for Unilever products, but it has not given up all research in this field.

Both U.S. and European labs are interested in polyunsaturates in oil for combating heart disease; major efforts have been mounted at Vlaardingen and Colworth on vegetable protein. Despite this extensive research in Europe, there is no joint project in this research between the U.S. and the European labs. The U.S. labs have maintained an orientation toward product development, whereas the European Central labs have also emphasized basic inquiry.

There are no joint programs between India and the lab at Vlaardingen but Hindustan-Lever has requested assistance on several projects, such as hardening of fats. The Vlaardingen lab prefers to send experts so that communication on complex problems can be more effective. The Indian lab was also invited by the Research Division to participate in the worldwide program in skin; it was interested in lightening skin, Isleworth in the tanning processes.

The Indian lab has requested help from Vlaardingen, partly as a result of which the Indian lab has built up its own capability in background research, and has reached the status of a true R&D unit, comparable to those in the United States. It therefore does not now need day-to-day assistance, though it can obtain it readily by simply asking.

Vlaardingen has an Information Department with sixty staff members including translators. It puts together the "Research News," which goes out automatically to all labs, and sends out reports of interest automatically to the information officer at other labs. Reports from other labs are received in this department, but there is little evidence in the reports of the Vlaardingen scientists of their having read reports from the lab in Bombay. Of ninety reports from India recorded in the 1974 KWAC reference index, only ten cited European and American lab reports, while citing twenty-four prior reports: three reports from Edgewater, fifteen from PS, and six from Vlaardingen/Duiven. Only four European reports cited Indian reports: of 558 reports from Vlaardingen/ Duiven, one cited two Indian reports; of 470 reports from PS, two cited seven Indian reports, and of 186 Isleworth reports, one cited two Indian reports; 12 St. Denis reports showed no citations to India, nor did 271 from Hamburg, or 539 from Colworth.

The department also has data bank terminals connected to the Lockheed service and that in the European Space Agency's research unit (ESA-ESRO), the cost of which exceeds $30,000 per year. The costs of the publications bought by the department is over $100,000 a year, covering some 900 journals plus books and governmental reports.

Finally, the information department has an editorial board that edits and translates research pieces for publication in Dutch, German, French, English, and U.S. journals. The rejection rate for such publications is only about 2 percent per year. Most of these publications come out of basic research rather than process research, since basic scientists have more time and are encouraged to publish; they recognize a personal advantage in doing so. Sixty-five articles in 1974 and seventy-two in 1975, were published by Vlaardingen/ Duiven professionals. The drive to publish in the middle and later stages of research comes from public relations and marketing (essentially on evaluation and safety testing) to improve the reception of the products, therefore reaching a quite different audience from articles on fundamental research.[6] Consequently, one finds a skewed distribution of publication across the sections of the lab. Even so, there is evidence that the publications are reaching large audiences, since frequent requests for reprints are received.

Despite extensive reporting, communication remains difficult. For example, Vlaardingen was doing research on fatty acids in the body's enzyme system that suppressed metabolism of prostaglandin development (not normally a good thing, though it might be in the case of an inflammation, and possibly in sunburn). Scientists at the

St. Denis lab in France read the report on this project but mis-interpreted it and reported negative results on some tests requested by Vlaardingen. Later, after a visit by a Vlaardingen scientist, St. Denis got the information straight and obtained more positive test results. Without personal communication, the results of the effort would have been different. When details are needed, the experience at Unilever is that a personal visit is necessary.

Assistance to Development Labs. An R&D Applications Unit (RDAU) has been set up at Vlaardingen for the purpose of support-ing development work and troubleshooting in six of the smaller European countries. The need for such a unit was first seen in the detergents area, where various factories could not afford to develop their own lab facilities. Some of the companies need lab facilities close by, for quick checking, and the RDAU is an effort to provide them with prompt service from a centralized source. The staff at the RDAU comes from various European nations, providing appropriate languages.

The RDAU was established by cutting Vlaardingen's development lab in half, leaving all of the development personnel that were in the small companies but shifting their roles to problem identification and to the implementation and modification of solutions provided by the RDAU. The RDAU was established because the small companies could not afford to develop their own labs that were necessarily less efficient. The smaller companies did not want to subcontract their research problems to a larger affiliated company, and the latter were not interested in taking on such work anyway.

The idea for the RDAU originated after a Port Sunlight section manager was assigned to Spain for two years. He found that plant development managers received reports of the Technical Information Service and could understand them, but did not have time to read all of them, much less check out possible uses of the information. In-stead, they jumped into their own developmental work with about half of the information they needed and therefore proceeded incom-pletely. The RDAU was seen as necessary also to centralize testing for the regulatory agencies.

The heads of the affiliated companies could see the advantage of this centralization fairly readily, but the heads of the development labs saw their own responsibility being substantially reduced. How-ever, location of developmental research at the small operating companies required a distribution of proper information and its proper use beyond the capacity of the units at these companies.

The RDAU is also involved in the consumer/market research done by the smaller affiliates in that its own evaluation work has to reflect consumer needs as well as market success. It is the responsibility of the local companies to generate appropriate consumer research, but they have not always done it, giving preference to market research.

The development of the RDAU was possible because of its limitation to Europe, where any plant can be reached overnight by rail or air. The assistance is also more readily accepted since the technician sent is frequently a member of the same company, attached to RDAU, and because there is a growing cultural unity in Europe. However, it would appear difficult to develop an RDAU for overseas, unless each regional area had a separate one. The company has not reached this point in development, yet it cannot put a detergent lab in each country.

The lab at Vlaardingen does some research in agricultural products that would be useful to developing countries, such as in oil seed crops and vegetable oils. It has liaison with labs in Turkey, Kenya, Brazil, Guyana, and India on sunflower seeds. With Turkey, it has provided an in-depth evaluation of the rape seed program (an alternative to sunflowers because they are less readily diseased), as well as on tobacco seeds and various other oil seeds such as poppies.

The lab in Turkey has not yet developed to an affiliated status simply because it is not big enough. Its sunflower program was undertaken through government sponsorship; it often requests assistance from Duiven, which makes a desk analysis of the field research reported to them and compares it to that in India on sunflowers and other oil seeds, sending the results of its analysis to both Turkey and India. Vlaardingen has investigated how good the sunflower is as a seed source, how important it is to the farmer, and how it fits with his other crops. Sunflowers have been grown on an experimental basis there, and the lab keeps tabs on the experiments and feeds Unilever assessments back to Turkey.

Unilever similarly sent data to assist the Mexican government in restarting a sunflower project that had been dropped. Vlaardingen also works with Brazil on similar projects, and has a project on safflower seeds, which have a higher level of polyunsaturated oils.

The ready flow of information is of the greatest importance and value, and some officials feel it is undervalued within Unilever in the sense that affiliates are not required to pay their full share. As one of many examples, a technical director in one of the South American subsidiaries wanted to develop dried soups, and requested information from Colworth on how to go about it. Colworth officials sent a five-page letter and four reports which put him into the business effectively—a package worth well over $100,000.

AFFILIATE LABS-U.S.

The R&D labs in the United States are located at Edgewater, N.J. (attached to Lever Bros.) and at Englewood Cliffs, N.J. (attached to Lipton); both are affiliated with the Research Division and do undertake fundamental research, thus distinguishing them from company development labs. The "research" part of the activities of these labs is organized into Scientific Research, Technical Services, and Administrative Services departments that support the operating divisions. Scientific research undertaken is both basic and applied, though probably 95 percent is for direct application to the business units.

Since the primary focus of this study is one the relationships between labs at headquarters and those in overseas countries within the multinational enterprise, a major part of the interviews examined the extent and nature of coordination among the labs. In the case of Lever Bros., there is a history of great caution in avoiding excessive management control from overseas that has tended to limit cooperation between Unilever (U.K.) and Lever Bros. (U.S.). Lever Bros. has found that fitting segments of R&D projects together across the ocean is a very difficult process. Consequently, some 90 percent of the activities of Lever Bros. R&D is not significantly different in scope or operations from similar labs in competing U.S. companies, with some 10 percent in cooperative research.

Lever Bros. Lab

The research at the Lever Bros. lab (Edgewater) reflects the product-orientation of the company to foods and personal and household products. Organizationally, Edgewater is headed by a vice president for research, who reports to the president of Lever Bros. (U.S.). He reports also on a dotted-line basis to the head of the Research Division in London; and is a visiting member of the relevant Research Planning Groups, which control 90 percent of the product research budgets. It is through these RPGs that the U.S. lab gets work done for it by the European labs.

Reflecting the ties to the operating company, there is a director of development for each of the major lines—household products, personal products, and foods—who reports to the vice president/general manager of each of these lines. But these directors also report on a dotted-line basis to the vice president of research for Lever Bros. They are responsible, within the R&D program of the lab, for looking after the development needs of their product area, and development managers of specific products report to them.

The Director of Scientific Research in the lab is responsible for biochemistry and microbiology, organic chemistry and patent coordination, perfumes and flavor, physical chemistry, and toxicology. The Director of Technical Services has managers reporting to him in the areas of analytical chemistry, manufacturing standards, packaging services, process engineering, and product and idea testing. In addition, there is a director of administrative services with responsibility for budget, engineering services, mechanical supervision, operation services, personnel, and other administrative services.

Program. The primary mission of the lab is for new product development; the second, the restaging of existing products; and the third, product and process improvements to make them more profitable and safe for the consumer and the environment.

Initiatives on any project may come from the departments or individual professionals within the labs; if other departments in the lab need to be involved, there is a procedure for approval by each department director and the Vice President of Research. Funds may be taken from two or three sectors of the lab budget for the project, and approval given without referral to the U.K. Research Division.

The scope of each program is determined by the market and local regulations in the country concerning components, packaging, etc. According to lab officials, the R&D effort is a function of the size of the market and its composition, the percentage penetration in the market, the nature of the competition, and the opportunity for introduction of new products. Programs required by the market would average about 50 percent of the research effort on the typical product, the need to comply with government regulation dictating the other 50 percent. This last percentage, however, varies with each product, and sometimes consumes 80 percent of research time.

Research programs in the United States tend to be locally-generated. Seldom is it the case that a manufacturing unit can pick up a product originating in another country and adopt it unchanged; it always needs adaptation in packaging, flavor consistency, shelf life, or some other area. Thus, the personal products group has attempted to introduce shampoos, antiperspirants, and mouthwashes from the United States, but to no avail. Market tastes tend to be significantly different—for example, Germans like a medicinal flavor in any oral product.

Sometimes, however, one finds an interesting two-way flow of information, as with a hard-surface cleaning product developed in the United Kingdom from an experimental product originating in the United States. Early market tests in the United States were not successful, because of the product's performance and packaging. But the United Kingdom picked up the product, altered it, and it is now being retested in the United States to determine whether it is a feasible item. There are also, of course, some striking examples of products successfully developed for the U.S. market that have subsequently been transferred elsewhere in the world—for instance, "Close-Up" and "Stripe" toothpaste (sold internationally as "Signal").

Coordination with Unilever-U.K. Policy coordination is achieved through the Research Planning Group under each of the product Coordinators. The Lever Bros. Research Vice President has a dotted-line relationship to these RPGs and attends their meetings as an associate. He will generally take to the RPG proposals on what the other labs in the system might do *for* the U.S. labs, rather than *on* U.S. projects. In other words, the U.S. lab might conceive of something that a U.K. lab could do for it, but it would not ask the U.K. lab to take on a major segment of an existing research project. Although the U.K. labs have multinational, multidivisional responsibility, the U.S. labs are multinational only to the extent that they do work for Canada.

The problem in trying to get a U.K. lab to do a piece of a U.S. project is that the U.K. professionals would not carry it out the way the U.S. scientists would wish. Tension arises among them, because scientists are not as cosmopolitan as they are assumed to be. Rather, Lever officials state, scientists are parochial in their understanding of the way in which products are used, thinking only of national market uses. There are further distinctive problems faced by the U.S. company quite apart from differences in product use. U.S. government regulations concerning safety and proof of product efficiency are generally more complex than those in other countries. In addition, shelf life, storage conditions, and competition all introduce especially stringent requirements into the marketing of U.S. products.

Nevertheless, central lab contributions can often be significant. What is most helpful for Edgewater is for the U.K. lab to undertake a particular piece of research that the U.S. lab is not prepared to do. The U.S. lab will then specify exactly what it wants checked out, and the U.K. lab will undertake to do it. So far, there has been no charge for such work; if there were, Edgewater would likely tool up to do it itself, for there is a considerable cost in time and effort in adapting a project to cooperative needs.

However, some projects are best done cooperatively. One was the development of deodorants for soaps. Another was the testing of detergent additives for environmental safety, an outstanding example of cooperation in which the needs of the U.S. lab were strongly backed up by the excellent facilities and techniques of Port Sunlight and Colworth.

In any such joint project, however, one of the major problems is that of monitoring communication. Written reports are passed back and forth, as are letters, but much is done by telephone (on a personal, old-boy basis), each checking the other's processes and making suggestions. Following through these suggestions takes considerable time, but without the telephone, the two teams would not be stimulated to amplify or modify their procedures. The telephone is used also as a means to force written amplification so that the record is complete, and is most effective when personal ties have been established through visits.

The personal ties, which are necessary to carry out effective communication, are developed out of longevity and regular visits between the labs. Some 350 professional will come from Europe to visit per year throughout the U.S., with 50 to 60 visiting Lever Bros. for specific purposes. However, only about thirty of the professionals at Edgewater have personal ties with their opposite numbers in the U.K. Officials indicated that this is probably enough, and that these ties will be reinforced by trips every two to three years in which a professional will visit between twenty and thirty people in the U.K.

The first step in cooperation, of course, is recognition of mutual interests and of complementary capabilities. This recognition is facilitated by the reports from the U.K. labs, which are read by each appropriate section in order to determine (a) to do nothing, (b) adapt information to their own projects, (c) suggest modifications in U.K. work for U.S. purposes, (d) suggest better techniques or methods from U.S. experience. If the last was desirable, U.S. scientists would notify their U.K. counterparts of something being done in the U.S. that might be useful in the United Kingdom. To give an idea of the extent of the exchange of reports, the Edgewater lab received some 250 reports unrequested from the U.K. labs in the first half of 1976. It received another 200 that it had requested previously, and had sent requests for some 250, some of which were received in this period with others yet to be received. In addition, it received some 150 patents it had requested.

The director of development in foods at Edgewater has most of his contacts with Vlaardingen, since the primary product line there is edible fats. He noted that Vlaardingen has an interest in what is done in the U.S. lab and will help if it can on any U.S. project. Visits occur at least twice a year each way, but the development work in Holland is only marginally useful to the United States because of differences in consumer tastes and distribution conditions. In addition, the U.S. Food and Drug Administration's recipes for food products are like a cookbook with very narrow limits, as in mayonnaise; the independence between the development departments in food in the U.S. and in Holland is 60 to 70 percent a result of market differences and 30 to 40 percent a result of FDA regulations. The restrictiveness of FDA regulations is easing somewhat, since these percentages would have been reversed ten years ago. The FDA is relying increasingly on international standards as a result of the U.S. government's desire to get U.S. products into world markets. Consequently, the Canadian subsidiary of Unilever leans on the United States in food development, and the United States in turn relies on the United Kingdom for some of its developments.

Lipton Lab

Since 1960, most of Lipton's research has been in tea and new products—on flavors and developing instant forms of foods. Some of the tea research is done in horticultural experiments in South Carolina, and some protein research on foods for high nutrition is done in cooperation with the United Nations and other world bodies.

As with Edgewater, this lab plays a quality control and assurance role in the plant. It is, therefore, concerned with improvement of processes, determination of appropriate standards, creation of better machinery and electronic controls, and the like. Quality control in the food industry is still in its infancy around the world—even in Europe—in the view of lab officials. Governmental pressure has still not raised these controls for all companies to the levels that are necessary and desirable. Therefore, much R&D will be related to this problem for some time to come.

Allocation of the R&D budget to various projects is largely done by "need," which reflects the political pressures among the product divisions, itself reflecting the profit pictures, the determined need for new products, and the persuasiveness of the various officials. Determination of where new products are needed comes from the New Products Control Committee made up of top management in the company (at the vice president level) in consultation with the New Products Coordination Committee, which seeks to develop products in a logical sequence. Once these groups establish technical feasibliity, market opportunity, and the "fit" in the product line, an exploratory research project is set up, taking one or two months to develop prototype and project costs. R&D has complete control of all development activities until such time as the product is ready for a plant trial. At that time, assuming successful implementation of the process, the accountability is moved to manufacturing.

The research budget is limited to one percent of total sales. If the lab has a sizeable new proposal, however, it might push for a larger budget, but it has not done so for several years. The budget is approved ahead of time on an estimate of cost and is charged to "research" on each product line; each division in the manufacturing company thus pays its proportionate share of the research budget, according to its sales. Even so, there is an annual review of each five-year project and new proposals made from within the lab. The scope of the lab's activities is strictly controlled by a policy that they should be directly related to the possibilities of commercialization of any developments.

Program. The quality control and assurance function has become so important at Lipton that U.K. labs are asking it for data on its testing and control procedures. In 1960, Lipton had one lab employee concerned with quality control. A statistical quality control individual was added (because Lever Bros. had one), and in 1961 the function was moved to the plant. In 1973, a study made by Booz-Allen-Hamilton determined that some of the quality control problems at the plant resulted from the "judge and jury" being in the same unit. This meant that quality control was sacrificed to the need to cut costs, generating a lack of concern by the worker for quality and productivity. The pressure was, in fact, against quality control and high standards. The consulting firm recommended the shift of quality control back to the lab. Around the same time, a publicized case concerning Good Humor ice cream shook the company up, because that particular division had been charged with selling its product below the legal standards on quality.

The consulting firm also recommended that greater attention be given quality control. With the move of the unit into the lab, the Vice President of Research has spent virtually all of his time in this area, despite only 50 percent of the lab's total activity being dedicated to quality control, process research, and quality assurance. Of this 50 percent, 20 percent is dictated by the necessity to meet FDA standards.

The major research in tea growing takes place in South Carolina. Tea was grown on a plantation at Charleston in 1912, but later was left unattended and was sold; the tea plants continued to grow wild. The plantation was bought back in the 1950s, and the company got help from Clemson University in South Carolina. The plants were then transplanted to an island, where the company is now growing select types of tea. They have developed mechanical harvesting, but the farmers, who are conservative, do not wish to expand tea growing in the area or do contract farming.

Another project has demonstrated that tea can be grown in California, where farmers would have to invest four years of time in developing the plants before they are mature. Consultants from the University of California at Davis had to learn all of Lipton's mistakes, but they are now able to assist in such an agricultural development. The knowledge is now simply "mothballed," ready to be used to substitute for a loss of foreign supply if necessary.

Exchange of Information and Cooperative Projects. In the food area, Lipton has acted autonomously without consulting Unilever, and at an early date tried to get into all aspects of foods—fats, carbohydrates, and protein. More recently it has restricted itself to tea and proteins, even though Unilever has done more work on proteins than perhaps anyone else in the world. Unilever has moved in and out of protein research, settling on its present interest in vegetable proteins and fish. (Lipton has been predicting the ascendancy of vegetable proteins over meats, but they have never quite made it; consequently, Lipton has lost money on the bet.)

Lipton has also done considerable work on gelatins, showing in experiments at Saint Luke's Hospital (N.Y.) with medical patients, that gelatin has health-giving qualities. Unilever Research has subsequently proved of some assistance in the Lipton gelatin products effort in the form of concept consulting and testing in an effort to substantiate potential product claims.

Some cross-exchange of information with Unilever was begun about twelve years ago on tea, after Unilever acquired the Lipton business outside North America. A Lipton researcher was looking through some Unilever research abstracts and found a report on removal of tea stains in fabrics, resulting from some research on detergent problems. He looked further and found two reports on tea chemistry related to the same problem. At the suggestion of Lipton officials, a British chemist visited the United States, building long-standing relationships and causing some tea research to be started in England.

In acquiring Lipton, Unilever took over a Lipton-held plantation in Sri Lanka (Ceylon), but this is now owned by the Sri Lanka government following its general takeover of estates. Tea research between Lipton and Unilever is now coordinated by a joint task force and reviewed on a two-year basis. Projects are allocated on the basis of capacity in each lab, with each providing its part cost-free to the other. Scientific and marketing personnel are part of the Tea Innovation Group. Unilever also owns another Sri Lanka company, Ceytea Ltd., which produces instant tea for sale on the world market.

Colworth has a joint project with Lipton's United States lab on Ceytea, with both providing technical research and doing technical work on green leaf tea in an effort to produce instant green tea; the operation in Sri Lanka has only limited technical expertise to do this research and to experiment in processing. The government there has let six to eight Unilever technicians come in for six weeks or more to help train Sri Lanka technicians, and Lipton has sent three technicians also to help train locals. The shared research projects put Lipton more into direct tea research and Unilever on indirect research supporting that of Lipton.

In vegetable proteins, Lipton has developed techniques for meat analogs. It commissioned Unilever to develop some of them; Unilever shifted some of its research operations, at no charge to Lipton, conducting the research as they saw fit, but under goals set by Lipton. This project continued for two years with substantial exchange of information, but did not produce a marketable product, and Lipton's goals changed in the process. This experience left some hard feelings among some of the professionals, since Lipton officials sometimes felt that Unilever was not heeding their desires.

Although Lipton does not often ask Unilever to undertake projects for it, it does so on specialities at times, such as cat food. Also, Unilever has the best lab in the world on toxicity of foods at Colworth—so good that the U.S. FDA has turned to it on several occasions for testing. Lipton would go quickly to the head of the Research Division in London if it needed assistance in this area.

Coordination with the Central labs is also accomplished through a substantial exchange of reports. Lipton receives the same types of reports that Lever Bros. does, in the fields of its own interests. All reports from the U.K. lab, plus those going into the TIS that come to Lipton, go to Lipton's Document Center for circulation within the lab. Beyond this, the lab may receive direct information from a researcher on a special project that he recognizes would be of interest. And data on any "hot" project is spread through the product Coordinators. In return, Lipton writes reports on new products, but not on product improvements or processes. All tea research reports go to

Unilever as well. Lipton tries to follow Unilever's system in reporting but it has cut back on new product reporting because of the cost.

The value of this exchange of classified and unclassified documents is large but subtle. For example, the projects at Colworth dovetail fairly well with those at Lipton on food safety. Lipton has, therefore, asked Unilever to make minor adjustments in their programs to help it out. Where Unilever has developed a particular methodology, it may adjust it at some stage in order to complete an experiment begun at Lipton. In addition, negative information supplied in the experiments on food safety is highly valuable to Lipton.

The large size of U.K. labs is of critical importance; smaller labs simply could not provide the assistance that Lipton needs at times. The extent and value of coordination, therefore, depends among other things upon the size of the parent lab, its choice of projects, the quality of its manpower and so forth, which are in turn determined by the market needs. Similarly, the nature of the programs at Lipton is determined by the size of its market, the products made by the manufacturing company, and the contribution it can obtain from Unilever, which equals at least 25 percent of the present Lipton research budget—that is, without Unilever backup, Lipton would have to expand its research by at least 25 percent.

Despite the support from the U.K. labs, there is seldom a full exploitation by Lipton of Unilever R&D; rather, it is drawn from a project here and there. Even when Lipton picks up a product, adaptation is required. Once the product has been adopted, Lipton will duplicate the work at the product development level to make certain that it understands the product thoroughly and has adapted it appropriately to the market. The evidence of its efforts to adapt products from Unilever is shown in the cost of unsuccessful attempts, amounting to $100,000 over two years.

Information used from "pertinent" reports was estimated by some of the lab officials to equal about 2 percent of the information flow. But this estimate should be increased by data on negative results, raising the value to 10 percent as a result of increased efficiency expanding resources by cutting off unnecessary effort.

Despite this coordination and support, and the wide exchange of information, communication remains difficult simply because of distance and the cost of closing it by telephone and personal visits; the latter are needed to close the differences in behavior and orientation of scientists. Marketing and technical representatives visit Lipton two or three times a year and Lipton personnel visit the Central Research labs.

Unilever sends some thirty on research programs and Lipton sends some twenty a year to the U.K. In marketing and manufacturing aspects, some fifteen to twenty visits are made each year both ways. These visits are approved by the Director to promote a constant flow of information. The joint project in tea calls for at least two visits per year; that on proteins required visits once or twice a year. Conferences of specialists on particular topics are provided by Unilever twice a year, but Lipton cannot get to all of these. In addition, every few months Lipton research managers visit Unilever labs in order to get up-to-date review of their programs. The Director goes to conferences of his peers, and as an ex officio member of the RPGs is asked to attend once a month, though he does not always make it. On the tea project, the section head in Lipton has spent a full year in Sri Lanka and three months in the U.K., and returns every other year or so for shorter visits.

The Canadian operation takes off from U.S. R&D, but adapts the products to its own market. Since Lipton is in basic research in instant tea, the Canadian company does not have to duplicate this, merely performing some product development or improvement. Though it might appear that the Canadian lab would not have to make much adaptation in Lipton products, there are in fact differences in the Canadian market's tastes and in the way in which the products need to be presented. This close relationship can exist only because of the physical proximity, reducing travel costs. (Trips and telephone calls to the U.K. are carefully screened by Lipton, though they are not if the trip is to California or to Canada.) Lipton charges Canada for the direct time and service to the Canadian lab. Exchanges of personnel occur in marketing and research, though there is no formal channel for periodic information transfer.

Contacts with Science Community. The availability of contacts with university personnel in the same scientific areas is significant to the continuing success of the lab. The contracts for research with nearby Rutgers University include enzyme reactions of tea and the problems of insoluble enzymes. The techniques developed by the Rutgers group are unique. Though the central labs might well have developed the same approaches, Lipton is very careful to rely on them only for high-priority projects in which Unilever has both a high capacity and strong interest. Much of what Lipton gets from the university labs could be gotten from Unilever, but they do not do so because of their caution to remain as independent of Unilever as possible. Lipton does not contribute to central R&D funding, and therefore does not wish to take much out of it that is specially provided to them, as contrasted with what is generally provided to all labs through the media of reporting and visits.

Consequently, Lipton hired consultants on process technology, since Unilever did not appear to have that expertise. Contracts exist on nutrition with MIT professors, with University of Minnesota professors on food processing, with Columbia University professors on metabolic testing, with Georgetown Medical School, with various hospital labs, and with Clemson (University) and the University of California on tea. The cat food project includes contracts with professors at Rutgers and the University of California.

Suppliers' interest in improving their products also increases the research available to Lipton. Research by suppliers is especially helpful in protein (meat-extenders). The same can be said for flavors (chicken, roast beef, and pork flavors in soups) where the companies are putting 10 percent of their sales in research. This reduces the cost of research to Lipton.

In addition to these support facilities, Lipton relies on U.S. Department of Agriculture research in its fields, such as dehydration. The U.S. Agricultural Extension Service is another basic foundation of food R&D in the country, not only providing new food varieties but stimulating demand. And the agricultural services arising out of the U.S. land-grant colleges provide strong support for Lipton's own R&D program.

AFFILIATE LAB-INDIA

The research lab in Bombay, India is part of Hindustan-Lever Ltd. (HLL). It reports to the Research Division, making it an affiliated lab. Its purpose has been to work closely with HLL in evolving processes to upgrade indigenous raw materials as part of an import substitution program and in improving the product line. The R&D center has made it possible for HLL to use several unconventional oils, e.g., castor, ricebran, sal, etc., in soapmaking. With the lab's help, the product line expanded from soaps and vegetable ghee to detergents to convenience foods, dairy products, and animal feeds by 1960.

HLL went through several changes prior to 1960. It was established as a trading company in 1883, owned 100 percent by Unilever. By 1930 it was marketing essentially Unilever products; it entered manufacturing in the early 1930s, and by 1955 it was developing some of its own technology. In 1956, Unilever sold 10 percent of the equity to Indian shareholders, raising it to 15 percent in 1965. (Unilever is currently in the process of further raising local shareholdings.)

By 1961 HLL was managed largely by Indians with an Indian chairman; by 1966, its management, technical, and developmental personnel were entirely Indian. In this process, control by Unilever has been relaxed to where it now reserves for itself only the approval of capital expenditures over a certain amount and promotions above a certain level. It plays an advisory role in other areas, but the company is essentially run autonomously. This autonomy permits the Indian managers and the company to maintain a loyalty primarily to Indian interests. Their careers are Indian-oriented, rather than Unilever-oriented; several of the managers have left for the public sector, such as state trading companies, steel, and chemical companies, as a result of a call from the government for high-level managers. Government officials have seen HLL as a source of professional management with loyalty to India. (This is reportedly different from many other foreign-owned affiliates in India, whose managers see themselves as British or even American, have foreign lifestyles, and act more foreign than the foreigners themselves.) Despite this national loyalty of management, HLL has not contributed to political parties, though it is normal for both corporations and corporate officials to do so. It has sought to keep an impartial image in government councils by remaining apart from such direct participation.

Origins and Organization of R&D Lab

The Vice Chairman of HLL determined in 1952 that the company should have an R&D lab, under the assumption that sales and products would expand fairly rapidly in India. After a study of the national oils and fats situation, HLL determined in 1958 that there would be a substantial shortage of these materials in India in the 1960s. A speech by the Chairman in April of 1959 looked forward to research as a means of increasing the yield of local oil seeds per acre and extending the new techniques to the village level. Because of a projected shortage of foreign exchange, HLL management foresaw a need to substitute indigenous products for imported items—diversifying the sources of oil seeds from cottonseed to oil palm and to minor oil seeds.

Although research had begun in perfumery chemicals in the late 1950s, and some research on oils and fats had been started in 1958, it was not until the introduction of convenience foods that the lab was really instituted as a separate part of HLL. The work of

European labs on dehydration and dairy products helped in setting up these lines. The company experimented with cheese, reflecting the capabilities of Dutch-Unilever, and lost its shirt because the fermentation "bug" was not effective and stable in the production phase, even though it had been in lab experiments. It went into dehydrated soup and sold only twenty tons of product, taking back sixteen tons that had stayed on the shelves too long, because soup is not a part of the normal Indian diet. This era reflected an effort by the R&D lab to dictate what should be produced and sold—a mistake from which the company learned to be more consumer-oriented. (Previously, new lab products were approved without much consumer research.)

Difficulties in the coordination between the R&D lab and the manufacturing company were illustrated by the company's move into nonsoap detergents (NSDs). It adapted NSD bar work done in Unilever companies in South Africa, Portugal, and Australia—where the detergent bar had not been successful. The work was carried out in 1966-1967 by the Factory Development Department of HLL, which modified the formulation in order to reduce sogginess, and then produced it very carefully on old soap processing equipment. The bar is one of HLL's best products, and the lab has since taken up supporting work in this area.

The problems between the lab and the manufacturing company stemmed partly from the lab's strong tie in its early years to the headquarters company in the U.K. through liaison officers, who took it upon themselves to support the R&D activity and to feed it potential projects from Unilever. The result was a tendency for the lab to work without clear commercial objectives related to local business. To reach its present success, the lab was gradually brought under tight control by the company through its Profit Centers; its programs have been woven into the project planning system, and it is required to account for its cost-effectiveness.

The contribution of Unilever to the development of the R&D lab was seen by HLL's present chairman as having the foresight to seize the opportunity to do R&D in India, largely because of Unilever's own experience and faith in R&D itself. Unilever saw the opportunity in (a) the peculiar problems of raw materials in India on which R&D was best done locally; (b) the market and HLL business being large enough to support R&D; and (c) the availability of qualified Indian scientists. Strengthening of the R&D lab has left HLL's competitors far behind. Unilever contributed by selecting the early scientists who had been educated at foreign universities, and then orienting them commercially at Unilever. Unilever also brought the

Indian lab into the international research system, helping it to learn background research and consumer-oriented R&D. More recently, Unilever has given continued encouragement to lab personnel, treating them with respect. The result has been that the lab has been elevated in status in both Unilever and in HLL; it now reports directly to the HLL chairman.

Despite HLL's autonomy of operations, HLL's chairman places a high value on continued links with Unilever R&D, which provides access to international R&D data and processes, training of scientists and exchange of personnel, training in R&D management, such as project planning and consumer research, and stimulation of professional curiosity through international contacts. Both the chairman of HLL and the head of Research visit with Unilever personnel at least two or three times a year. Personnel at all levels have found Unilever willing and eager to reply to inquiries and to extend assistance—at no expense to HLL.

Reporting to the head of the lab are four scientific divisions—biology, chemical physics, chemistry, and process—plus information and administration officers. The Biology Division undertakes work in biochemistry, microbiology nutrition, toxicology, and agricultural extension. The Chemical Physics Division is responsible for surface chemistry and catalysis. The Chemistry Division encompasses analytical chemistry, organic chemistry, perfumery, raw materials, toilet preparations, and basic chemicals. The Process Division covers engineering services, process engineering, process studies, and product assessment.

Some 80 percent of the work at the R&D lab is directly in support of project work for the manufacturing company. It therefore has achieved a close relationship with the factory in Bombay, which has a small development lab of its own. The development lab is linked to production, while the R&D lab has closer ties with marketing. When the R&D lab does develop a new product or process, however, it frequently puts a man in the factory to set up the equipment and check it out in a pilot operation.

Development Lab

The close relationship between the R&D lab and the development lab at the factory site arose from research's having been done on the factory site until 1967, when the new lab facilities were built. In 1957, when research was first begun with three to five professionals and ten support personnel under the chief chemist, it was exploring the opportunities for use of indigenous materials. It expanded in 1958 with work on perfumery chemicals and unconventional oils and fats as the major fields.

Processes developed at the Research Center in these fields were transferred into the plant where they were converted into economically viable processes by more technically-oriented personnel. There was a separate development manager with a few men working closely with the various production managers and process engineers. A clear need arose for a larger development group at the plant, and this was begun in 1968. The present factory manager (who holds a Ph.D from the University of Illinois) came out of this group, having been sent to Vlaardingen as head of the protein section for a year; he then spent two years as Production Manager before assuming his present position.

Reporting to him is the Development group in the factory, with fifty-two staff members. Reporting to the development manager are section heads responsible for soaps, detergents, oils and fats, chemical engineering, and process engineering. One of these has a Ph.D. and the rest B.S. degrees. Two department heads, one assistant department head, twelve technical assistants, and thirty workers make up the lab's staff. The largest demand for trained personnel in this group is for chemical engineers, who are needed to translate R&D output of the company lab into production processes. These engineers are able to work either in developmental labs or in production. The numbers of such trained professionals are increasing, and HLL can have its pick.

The upgrading in the educational level of personnel at the factory has had an important impact on the role of the development lab as well as on its relations with the R&D lab. Out of the 100 chemical engineers at the plant, 92 have a B.S. degree, and half of the 330 nonmanagement supervisors have B.S. degrees in either engineering or commerce. In addition, the hourly workers are increasingly obtaining high school degrees. This educational level compares to 70 percent illiteracy in 1950. The acceptance of change rises with education, and the spread of education has helped gain a higher level of technical competence and more rapid industrialization in the plant. Therefore, the output of the development and the R&D labs is more readily accepted in the plant.

Formal reports between the development lab and the R&D center concerning factory needs are made in a monthly review by the Technical Director of the profit center and the R&D lab head. If necessary, any problems are worked out between the head of the research lab and the factory manager himself. A second contact exists in quarterly reviews by HLL of R&D lab research that will be useful in marketing and manufacturing operations. Informal contacts exist between the development lab personnel and those of the R&D center.

The total developmental effort of HLL is divided between the factory's looking at the more immediate problems and the R&D center at those further away. Thus, immediate problems are 80 percent the responsibility of the factory development lab and 20 percent that of the R&D center, while any problems that are two years or more away are only 40 percent the responsibility of the development labs but 60 percent that of the R&D center. For example, castor oil processing was developed in the R&D center in a pilot plant, and the factory helped scale it up. R&D personnel may stay through the production scale-up or not, depending on the capacity of the development lab at the plant.

The program of the development lab in Bombay reflects the product line at the factory. It is responsible for all of the developments in the detergent field, and transfers information necessary to other plants in India through the Technical Director for detergents. It does the same thing for foods, though HLL's interest in this area is declining. It is also responsible for chemicals, in which the company is expanding; a new factory is likely to be built elsewhere in India for chemicals, with the developmental effort eventually spun off into it.

The Factory Development Department uses the Central Technical Information Service (TIS) to keep informed of new techniques in development. For example, when there was a serious problem of fungus growth on soap, which had been worked on at Port Sunlight in the late 1930s and subsequently, the factory found the reports through the computerized KWAC and was able to draw on them to help solve the problem. Later, Research was able to identify and characterize the fungal infestation and institute measures to eliminate the infestation from packaging material and "godowns" stocking the soap.

The manager asks for about 10 percent of the TIS abstracted reports listed, which may or may not turn out to be useful. Or he looks for specific reports on problems that have become intractable. The lab keeps no record of how frequently these reports have helped overcome problems, because they seem insignificant at the time, but life is made substantially easier for the production managers through "little solutions to little problems." For example, Unilever's work on catalysts has made it possible for the factory to produce "tailor-made" catalysts, which it could not have done otherwise; the factory at Bombay would not even have known of this work without the TIS. The TIS essentially disseminates information on available reports and helps to answer questions not yet asked or formulated by the engineers at Bombay. If these engineers had seen the problem and defined it, they could have simply asked for solutions. But merely reading through the TIS helps them to see that problems are in fact arising and how to characterize them. The development lab and the R&D center are especially helpful in interpreting or adapting results obtained from abroad relative to the problems of the production managers. Thus, the section heads will read the TIS and then ask development lab personnel to check the results. For example, the statement in one of the TIS reports that one plant is having problems on keeping purity of a product at 90 percent would show the reader that this level could be reached, compared to the 80 percent in his own. This would force the production manager to ask if he also could reach 90 percent.

Work on fatty acids, nutrition, and health conducted by Unilever research is drawn on with reference to different problems in Hindustan Lever. But HLL is becoming good enough in many of its R&D projects to stand alone. For example, the R&D center has done studies on infant nutrition, doing experiments on its own in hospitals, since other Unilever companies are not in baby foods.

Facilities at R&D Lab

The R&D lab in Bombay is the largest in consumer products in India, the second largest lab of any sort in private industry in India (next to Ciba-Geigy), and the largest that covers a number of diversified fields. The lab also has the only toxicology facilities for testing consumer products in India.

The sophisticated analytical facilities at the lab include an electron microscope and GLC amplification units and accessories, which are made at the lab itself. Infrared, ultraviolet, and nuclear magnetic resonance spectroscopy are also available at the R&D center. However, it must turn to the Unilever Central labs for mass spectra and optical rotary dispersion techniques.

As an indication of the facilities available, the placement cost of the equipment at the laboratory would amount to between £6 and £9 million at present inflated prices, though its depreciated value on the books amounts to only £1.5 million.

R&D lab's budget increased from £290,000 in 1970 to £490,000 in 1975 and to an estimated £630,000 in 1976. Of this total cost, roughly half was for salaries and wages. Depreciation of equipment accounted for roughly 15 percent; another 12 percent was spent on sponsored scientific research, books, and periodicals, laboratory purchases, consultants, etc. Administrative costs have annually run less than 10 percent of the budget.

The library budget alone amounts to £10,000 per year for books and periodicals. The journals come mostly from the United States and the United Kingdom, because of the ready use of the English language, with most of the rest coming from India, also in English. Only two journals are obtained in a foreign language, that being German.

Of the overall funds, approximately half was expended for projects relating to soaps and detergents, just under a third for chemicals, around 10 percent for toilet preparations, and the rest for foods and animal nutrition.

The capabilities of the personnel in the R&D lab are illustrated in that, in addition to the head of research, each of the division chiefs has a Ph.D., as do each of the section heads, save that in engineering services. Of the total staff of 215, 42 hold Ph.D. degrees. All of these come from universities in the United Kingdom and the United States. Many research assistants have Indian Ph.D.s; ten officials hold the Masters degree, 125 the Bachelors degree, and 40 are technicians. Their research capabilities are illustrated by one man alone having had some sixty publications. To facilitate continued training, there is an extended exchange of one senior professional between the R&D lab in Bombay and one of the central labs of Unilever each year.

R&D Program

Although the total research program is largely dictated by the medium-term production program, the scope and nature of the projects in the R&D lab have to shift more frequently than production does, in order to anticipate new products and market needs. Consequently, it has done research on foods, animal feeds, skin, hair, oils, nutrition, detergents, and toxicology. Future research programs will include inorganic chemicals, and new business opportunities, for example, synthetic fatty acids based on parafin obtained from the government refinery. (That the materials come from the government refinery means that the company will be dependent on the government's willingness to supply.) In addition, the research program examines competitors' products, and provides services to the factories related to such topics as instrumentation and safety requirements in production.

Foods and nutrition. HLL's research objectives in the food area include specialty fats, feed for animals raised for human consumption, and work on nutrition. The work in food technology and nutrition at the lab was begun first in dehydrated food, from which twenty new products were developed, many using special Indian preparations. To accomplish this required the development of new knowledge on Indian raw materials, their processing, the behavior of processed food under varying climatic conditions, and the preferences of Indian consumers. Out of these efforts came a new process for the production of a standardized high-quality ghee.

Nutritional research stemmed from the formulation of a milk-based baby food suited to Indian requirements; programs of clinical evaluation followed. Further efforts in nutrition led to a project for the fomulation of high-quality protein foods combining several materials. A protein food was developed that was equal or superior to milk protein, using only about 15 percent of milk protein in the total. This mix, fortified with minerals and vitamins, has been converted to inexpensive products. This project required a combination of biochemistry, nutrition, food technology, and product assessment, employing linear computer programming techniques to obtain the lowest-cost, high-nutrition formulas.

The formulas for Sukhadi, the low-cost, high-nutrition product for mass feeding by the government, were derived by putting into a computer the amino acid analyses of the proteins from various vegetable materials and mathematically determining which mixtures of seeds were equivalent to the amino acid spectrum of milk. Animal experiments were done on the nonprotein fraction (mostly starch) from each vegetable source to determine the latter's effect on the utilization of protein. By feeding into the computer the constraints of cost and the protein efficiency ratios for each individual material and for various combinations, a few formulas for highly nutritious protein concentrates were obtained to produce the Sukhadi.

For mass feeding programs, it is not only important to obtain appropriate proteins with high biological efficiency, but attention must also be paid to the availability of various ingredients and the acceptability of the final product in various regions of India. The lab research, therefore, has to be combined with information on material sources and consumer tastes, even for the purpose of mass feeding. The Sukhadi was so acceptable that the government has used these concentrates in its own nutrition programs, feeding five million people a day during the drought in the first half of 1973 with 350 tons of the material each day.

Skin and hair. The HLL lab and a central lab are working on skin lighteners and hair growth. The labs have cooperated closely at a scientific level for several years. Since many of the products will come into direct contact with skin and hair, the labs have sought to understand biological control of skin conditions, pigmentation, and hair growth. The lab has developed an understanding of the mechanisms of dispersion of melanin, the major pigment of skin, and is now able to alter the skin color.[7] This research has also led to the creation of a skin-lightening cream, "Fair and Lovely," which is now in the market, and to a better understanding of reactions to sunburn and tanning.

The research on hair is directed at the causes of male baldness. These studies involve ultramicroanalysis of elements in the hair bulbs, using sophisticated radioisotope techniques, as well as examination of testosterone metabolism in the hair bulbs. Close contacts are maintained with local dermatologists and plastic surgeons, with scientists in the other Unilever labs, and, through them, with scientists in university medical schools for dialogues on the progress of this work.

Vegetable oils. A major emphasis in the program is on oils and chemicals, as the lab searches for useful minor oils and various fine chemicals. Twenty-four fine chemicals have been adopted by HLL, of which only three came directly from Unilever. The lab also worked on a process of removing the toxic principals from cottonseed through the extraction of the oil, leaving the meal available for human use; it was used previously only in animal feed. It also succeeded in detoxifying some sal-oil cake by removing the tannin, whereas previously the cake could not even be fed to some animals. The lab's research in minor oils has also developed new compounds and new catalysts for selective hydrogenation, opening new product uses. In addition, the lab is seeking synthetic material in anticipation that inedible oils it is now using will be made edible and will then become scarce. To increase the supply of vegetable oils in the country, the lab has developed an agricultural extension service to work with agricultural universities in Punjab, Tamil Nadu, Andhra Pradesh, and Maharashtra to improve sunflower-growing methods, emphasizing the improvement of seed selection and planting.

The research on sunflowers is aimed at increasing the availability of edible oils in India. The government is now supporting soy bean production because of the high protein content, but sunflowers need much less water, and have two growing seasons during the year. Better seed varieties are being sought from around the world;

Vlaardingen provided forty different seed samples, which the lab sorted through and tested in different land areas in India. They are looking for a short-stemmed, full-headed plant; this will droop the head so that birds cannot readily pick out the seeds. The lab has persuaded some farmers to cultivate sunflowers as a marginal crop, guaranteeing a price equal to 90 percent of the groundnut price fo for the sunflower seeds.

Soaps and detergents. The work in detergents is aimed at seeking new actives from fats and rosin derivatives, investigating detergent builders and ion exchangers, and developing bleach products and substances with ash removal capabilities. The lab is also doing performance evaluation on detergents versus soaps. The research program in soaps includes not only the development of synthetic fatty acids, but also the use of rice-bran oil and a number of minor oils for soapmaking.

Chemicals. In chemicals, it has extensive projects on natural essential oils such as beta pinene, palmarosa, citronella, and on a variety of alcohols. Examination of new product lines comes from the projects in making glycerol from molasses, the degradation of cellulose, biosynthesis of fat, treatment of ossein waste, TSOP from phosphoric acid, and filter aids.

Toxicological studies require tests of the effects of different degrees of intake of the materials—subacute (high doses, short time), acute (ninety days), and chronic (two years)—involving oral and parenteral administration of the materials, backed up by special investigations in pharmacological, biochemical, hematological, and pathological effects. These studies help identify the toxic materials in meals from the sal and mowrah seeds and how it is possible, by modifying their chemical nature, to detoxify them. The procedures developed for detoxification in these seed meals could be employed in others to remove tannin or saponin. The detoxified meals could then be used in animal feed stuffs.

Similarly, sal fat was studied in a subacute toxicity dosage in rats and was found to produce no harmful effects, holding out the promise of its utilization as a cocoa butter substitute. This would considerably benefit rural areas where sal seed is collected. Research continues in this area, along with that in other minor oils from seeds of the mango, karanja, castor, sal, and mustard.

Argemone seeds resemble mustard seeds and occur as an inadvertent contamination or deliberate adulterant in mustard seeds. Argemone oil has been suspected of causing glaucoma and dropsy, sometimes on an epidemic scale. The lab has conducted a subacute toxicity study in rats and established the tolerance limits for levels of contamination of edible oils with argemone oil.

Soaps made with processed minor oil, such as modifications of castor oil, rice-bran oil, linseed oil, and watermelon seed oil have been studied for potential irritation to skin and eyes, with the hope of increasing their use if results are negative. A new material was developed at the lab as a substitute for coconut oil in soaps, but despite considerable commercial interest to the company it was given up because of demonstrated toxicity.

All of the tests have been facilitated by the use of a scanning electron microscope, which is the first of its kind in Bombay, helping to identify early morphological changes indicating possible toxicity. The instrument is also regularly employed for understanding micro-topography of various materials and their development into useful substitutes for imported materials.

Animal feeds. The work on animal feeds is divided into three areas—poultry feeds, cattle feeds, and detoxification and upgrading of minor oil cakes and other materials currently used as manure or fed in only small amounts of livestock.

Existing world knowledge allows the formulation of feeds for animals using raw materials normally required for human feed, but considerable research is necessary to make cheap feed without using human food while maintaining the present safety and nutritional benefits for the animals. For example, spoiled ground nuts or soy-beans can be detoxified and fed to animals. In addition, many potentially useful forest products are not used by the farmer because experience has shown him it will not fatten chickens, produce yellow-yoke eggs, or provide good milk yields. Untreated oil cake from castor or mowrah, for example, would probably kill his chickens in three or four days.

The lab has tried to identify the metabolic benefits of seeds in an experimental chicken house. Although some materials have a high nitrogen content or plenty of starch, the birds may not utilize the material adequately. This may be due to (a) the particular structure and low digestibility of protein and carbohydrates, (b) the presence of inhibitory materials such as tannins, which precipitate protein and can discolor eggs, and (c) the incorrect balance of the dietary components, including vitamins and mineral salts.

The same types of problems exist in cattle nutrition, but existing studies apply to cows living in temperate climates. Problems such as the effect of temperature on a metabolism of the cow, the nature of the activity of the rumen bacteria, and the effect of the physical properties of the food on digestibility are little researched. Moreover, there is insufficient information on the difference in composition of buffalo milk from that of cow's milk.

Research on animal feeds is tied to that on nutrition, concentrating on the feeding of chicken (broilers and layers) and of buffaloes in order to produce higher fat and protein contents than in cows. The research on the water buffalo involves tests on the digestion and utilization of food by the animal to determine the most efficient feeding patterns in order to increase milk production without losing the high fat content. The lab is relying on a substantial volume of similar research at Colworth on other animals, seeking to determine differences between them and the buffalo. One of the problems is how to utilize the efficiency of the buffalo in converting grass and straw, which is cheaper than the diet of cows, into milk.

The program involves not only research at the lab, but projects on the farm and experimentation with a few cooperative farmers who are using the feed supplied by the lab. If their present research is successful, they will formulate a better compound feed for the buffaloes.

Cooperation with government. In addition to these research projects, the lab is involved in a variety of activities related to governmental policies. It has responsibility for drawing patent specifications, making applications, and opposing competitors' patents. It must follow up any complaints on the use of products, such as animal feeds or soaps, sending scientists for surveillance and testing and instituting procedures and establishing facilities for quality control in various locations throughout India. The lab also deals with the Food and Drug Administration of the state of Maharashtra on labeling and scientific support of advertising claims. It participated in the deliberations of the Central Committee of Food Standards and its various subcommittees concerning the addition of flavor and coloring in margarine. And its officials participated in a number of committees of the Indian Council of Agricultural Research, the Indian Standards Institution, the Council of Scientific and Industrial Research, and other public agencies.

Cost/Benefit to HLL. In order to demonstrate the value of the work of the lab to the rest of the company, an effort has been made to calculate the "cost/benefits" of research. Estimates of cost savings in production were made for some of the detergent products showing how much was saved from use of new oils or processed developed in the lab, as well as for perfumes and flavors as a result of reformulation and development of synthetic substitutes for natural essential oils, from the development of fine chemicals, the use of new oils and processes in edible foods and fats, and in toilet preparations. The total cost/benefits calculated by the lab on these items amounted to 45 million rupees (£2.5 million) in 1974, excluding any benefits accruing from service work in toxicology, microbiology, or nutrition. These estimated savings compare with the total budget of the R&D lab in 1974 of Rs9 million (£500,000), meaning that five times the total research budget or more was recovered through cost savings. Many of the potential savings are not included in this estimate, and no value is calculated for increased sales.

Reliance On Unilever

Despite much of the lab's program being oriented to use of indigenous materials and thus unable to rely directly on research undertaken by Unilever Central labs, the latter provide a wealth of knowledge concerning various aspects of the Bombay programs. For example, a number of fat blends suitable for making tropical table margarine have been prepared relying on Unilever research, and a number of flavors for these margarines have been obtained from Vlaardingen.

The results of central lab research are transmitted to Bombay through various reports, averaging between 120 and 140 per year. In addition, Unilever research personnel conducted over twenty seminars at the HLL lab from 1971 through 1975. There are formal exchanges of personnel for three months each year between the lab and Vlaardingen.

Within the total research program, however, the lab in India does not expect to receive *direct* support from central research on individual projects, and there is no real dovetailing of programs between it and the central labs. For example, HLL skin research was drawn somewhat from biological research in melanin at Isleworth. There was no real competition between Isleworth and Bombay in this project, because they took different routes. But though each kept the other informed, criticizing its processes and tests, and scientists visited both ways, neither was there the complementariness found in a coordinated project.

The type of support received is illustrated by a Russian process for manufacture of synthetic fatty acid, previously examined by Unilever labs and turned down as unsuitable for Europe because of the lower quality of the product. The Central labs did not try to apply it to the Indian situation, but HLL picked it up despite its unsuitability for Europe. The Russians had overdesigned the process and obtained yields of only 50 percent from raw material; the lab successfully adapted it to a lower cost process with higher yields. From this research, the lab moved into a pilot plant operation that successfully eliminated the stages of thermal treatment and distillation, raising the yields to 80 percent and cutting costs in half.

Although it is difficult to quantify the reliance of the lab on Unilever labs and the gains it makes because of this association, the links to Unilever R&D have helped to initiate and to accelerate HLL's work in fatty acids, in microbiology, in fractionation of sal fat, in producing catalysts, in fistulation of animals, and in toxicology experiments. The training of personnel, the injection of ideas, and reduction of time in developing the various projects are significant contributions to the Indian lab's program. In addition, it has learned techniques of managing R&D and ways to improve scientific methodology and techniques, such as the use of mass-spectrometry and gas-liquid chromatography. Probably of greatest importance, however, is the training of scientists that not only makes them more efficient, but also eliminates the cost of education outside HLL.

Much of the contribution from Unilever comes through personal contacts regarding a particular project—for example, "Why don't you try this?" As a result of such communication, the HLL lab adapted techniques in the extraction and bleaching of minor oils, melding these results with continuous information from Unilever research reports. Then they checked the results on personal visits with Unilever scientists.

Researchers were interested in a sulfonated fatty acid process, that enables them to extend the availability of fatty matter for soapmaking. However, sulfonated fatty acids sensitized the skin. Unilever researchers advised them not to continue the project, but HLL found that rice-bran oil (which was cheap and available) could be sulfonated and incorporated into the soap bar. Then it drew on two-decade-old Unilever research on a completely different product and another project, two years old, to reduce the sensitization. The work of the Unilever researchers alerted HLL to some dangers, which permitted it to avoid over two years of work.

On another project, the lab was in correspondence with Edgewater on its work of ten years ago on a special chemical that may be explosive in processing. They have also taken the full-blown product, "Close Up," from Edgewater, and have used technical information from Isleworth in England in some products. HLL has also drawn heavily on Unilever with reference to detergents and hardening of fats.

The development of sal oil is an example of the extent to which the lab relies on Unilever research. That this particular seed produced a useful oil was not known before 1966; the glyceride structure of sal fat had been publicly known, but not seen as useful. Investigation revealed two potential uses—one in soap and the other as a substitute for cocoa butter. Different processes were required, and the cocoa butter use gives a better return, so the lab moved into the second. Through these stages, the lab at Welwyn (England) assisted by visits to the HLL lab. Samples of the fat were also sent to Welwyn, which found it to be satisfactory for use as it was. The company therefore exported some of it.

The research then turned to the changes in the chemical structure, showing that the fat was unstable. Experiments were made at both HLL and Welwyn, but not all of the answers have been found to date. Enough has been found to be able to sell the product, and some of the instability has been removed by refining. However, they have found that a minor component, say "P," constitutes 5 percent of the fat when first processed, but that when stored, component "P" shifts to "Q" and its percentage content changes. The lab at Welwyn is helping to determine the nature of "P" and "Q", for the existence of "Q" reduces the melting temperature of chocolate, eliminating the usefulness of the sal oil in chocolate. This investigation has had to go back to basics—the forest conditions (weather and growing environment), the collection process storage, effects of insects, and the changes introduced by the processing of the fat.

The lab has now fractionated sal fat to get the hard portion separated from the soft, using the former in chocolate and the latter in soap. It is now exporting a higher quality of sal in an amount between £3 and £4 million.

In addition to Unilever's assistance on particular projects, the scientists at HLL find that the company's "Patents Abstracts" are useful in signaling potential new products for India. The "Research News" sent out by Unilever is also read by all research scientists to determine which reports they wish to request.

The Information Division of the lab has responsibility for circulation of the "Research News" and reports to management staff. Each will mark items to be requested from the various labs throughout Unilever. Vlaardingen automatically sends all reports except those on foods; Port Sunlight also sends its reports in some fields automatically. The research reports, once requested, arrive in ten to twelve days, which is prompt enough for their effective use. During 1975, the lab requested 270 reports, and obtained 795 automatically. Other labs in Unilever requested 343 of HLL's reports. Of these,

HLL sent 53 to and received 73 from Colworth; sent 58 and requested 29 from Port Sunlight; sent 61 and requested 2 from Vlaardingen (in addition to the 332 received automatically); sent 38 to Edgewater and requested 6.

In writing their own research reports, HLL management requires that all use of reports from Port Sunlight and Vlaardingen must be cited; substantial references are made to reports of these labs, as well as to those from Hamburg and Isleworth. Most references are to older reports and to background research, showing that the research at HLL does not really dovetail or parallel that going on currently in the Central labs. Nor do the reports of the other labs cite HLL reports frequently, though they have had a few "best sellers," as in 1974 on spices, Indian medicinal products, and sal fats.

The chairman of HLL stressed that the Unilever research system is relied on to stimulate activity in the Bombay lab, since professionals realize that they belong to a wider system of scientific thought and knowledge out of which new ideas are coming continuously. The personal contacts with scientists of the other labs activates a search for new methods and approaches at HLL. This stimulus comes from the mere asking of questions, formulation of new hypotheses, and inducement to search for new ideas. The results are seen in the stimulation of projects, the avoidance of failures, the reduction of research time, the elimination of dead ends found by other labs, and thus the reduction of costs due to a higher percentage of successful experiments.

Contributions of R&D Lab

The contributions of the lab to HLL's success are seen in the lab's various programs—development of minor oils, fine chemicals, and new areas of business, plus product assessment and safety testing. These have given HLL a high standing with the government and the intelligentsia of India. They have also attracted better management for the company and better R&D professionals. Finally, of course, these activities have kept the company technically sophisticated and ahead of its competition. It is so far ahead that critics suggested that HLL should *give* its developments to competitors, so that they will sell better soap and not use imported oils, which are expensive and take scarce foreign exchange.

The lab also contributes significantly to the development of the Indian economy. For example, the development of sal oil has increased exports, and the use of unconventional oils for soapmaking has reduced imports of tallow or indigenous edible oils in soap-making.

The research in fats and oils has provided a significant import substitution for the country, as evidenced by Table B1.3, which shows the expanded use of a variety of local oils by Hindustan Lever Ltd. When domestic inflation threatened to undo the beneficial results on the balance of payments brought about by a governmental proposal to reduce the price of scarce oils and fats by permitting imports, HLL saw that this would slow the long-term development of domestic oils. It proposed to the government a limitation of peanut oils to 20 percent of total usage, with the rest coming from cottonseed oil and minor oils that were locally available.

In the case of tallow, the government decided to import to keep the price down; HLL recommended against this, even though they will make more money through use of low-cost tallow. The company thought it preferable to decontrol the price of fats and oils, thus promoting local production. The Prime Minister asked whether production would actually increase, and at what price levels; HLL gave assurances that prices would not rise above certain levels. It was able to do so partly by research on indigenous several minor oils, which added to the total oil resources. The company was eventually able to convince the government that continued tallow imports would hurt the Indian farmer and benefit the foreign exporter without increasing the supply of edible oils.

Similar efforts have been made in the field of chemicals, relying on Indian turpentine, citronella from Java (being grown under contract for HLL at a price half that of indigenous oils), and Indian lemon grass oil for perfume. HLL has developed four chemicals from lemon grass, five from citronella, one from turpentine, plus eight from petrochemicals, two from essential oils, and has taken four directly from Unilever labs. In addition, there are twenty-two other chemicals under development.

Many of the above contributions have provided new processes for use of indigenous raw materials, substituting for imports and increasing supplies for industries based on oils and fats:

processes for use of castor oils, rice-bran oil, Kusum oil, and Karanja oil in soapmaking, plus mowrah oil and linseed oil in toilet soaps

cottonseed oil for hydrogenated fats

nickel catalysts for fat hydrogenation

detoxification of sal extractions to permit use at higher levels in animal feeds

Table B1.3. Oil Usage by Hindustan Lever (Tons)

	1967	1968	1969	1970	1971	1972	1973	1974	1975
Tallow/Palm	45,151	40,388	31,834	24,357	24,574	24,738	15,119	13,134	6,087
Coconut	3,066	3,201	4,105	3,205	3,542	3,461	3,450	1,455	6,192
Groundnut	12,476	16,635	9,167	10,244	9,812	6,826	4,205	7,500	7,927
Castor	1,043	649	7,535	11,505	7,671	9,444	1,706	9,923	16,212
Linseed	444	4,800	7,909	5,408	2,373	7,614	9,535	1,993	8,545
Mowrah	842	7,549	9,424	4,411	6,737	10,244	5,733	3,041	13,246
Ricebran	527	2,462	3,102	8,024	10,359	10,141	8,763	11,241	13,274
Kusum	177	205	659	1,270	2,247	1,771	1,083	375	773
Neem	196	685	782	1,570	1,551	2,928	1,776	2,089	3,061
Karanja	21	44	422	907	818	431	—	358	136
Sal	—	—	—	386	1,728	1,934	4,731	1,817	2,738
Others	3,835	7,180	11,993	8,779	8,314	4,070	5,235	3,930	4,925
Rosin	5,048	1,730	1,941	1,631	2,886	2,577	2,410	1,163	2,402
Total:	72,826	85,528	88,873	81,697	82,612	86,179	62,746	58,519	85,518
Total unconventional oil	7,085	23,574	41,826	42,260	41,798	48,577	38,562	34,767	62,921
% of unconventional oil used	10.6	28.1	48.6	52.8	52.2	57.8	63.9	60.9	75.7

laboratory animal feed and dehydrated processed vegetables
a synthetic detergent bar, creating in India the world's largest market for the product
a number of fine chemicals and derivatives of indigenous essential oils
export of covo, a triglyceride used in cocoa butter manufacture (valued approximately at Rs 40 million)
linalool, a complicated fine chemical
thin film evaporator for glycerine distillation with higher yields
use of improved support material for nickel catalyst
process for very high reuse of nickel from spent catalyst
processes for manufacture of synthetic fatty acid from paraffin and glycerine from molasses

Taking all of these contributions into account, total foreign exchange savings during the period 1962-1975 have been estimated to be Rs 679 million, as shown in Table B1.4. (The annual savings rose from about £70,000 in 1962 at 1976 exchange rates, to over £7 million in 1975.)

Another research project in which HLL has engaged is the development of prawn farming. Exports of prawn have been significant in India's trade, representing about 4 percent of total exports. The catch of prawns on the West Coast has gradually decreased, indicating that the principal fishing grounds are overexploited. Experiments in controlling the environment of prawn growth from eggs to adults,

Table B1.4. Exchange Savings Realized by HLL Research

	Rs million[a]
Fine chemicals/essential oils	103
Manufacture of catalysts	70
Development & use of chemicals mainly in the area of lather boosters, chemicals for toilet preparations, detergents, colors and pigments	26
Development of nonsoap detergent bar to replace soap made from imported oils	114
Development of local oils and fats to replace imported	366
Total	679

[a]At exchange rates in December 1976, the total was equivalent to £46.4 million or $78.9 million.

through its six stages of development, indicate that it is possible to obtain about a thousand tons of exportable prawn per year from a two-hundred-hectare farm. The problem is that in each stage of its life cycle, the prawn requires a different kind of food. Furthermore, the food items at the later stages are also needed for human consumption (clams, shrimps, and fish), making it necessary to develop a synthetic food for prawns.

The HLL lab began experiments in growing prawns under controlled conditions in 1972. A compounded food was developed that produced better growth and greater survival in the later stages than fresh clam meat. While the experiments have been of great value in establishing familiarity in a new area of work and in understanding features such as feed needs and composition, further experiments will need to be carried out while simulating complex conditions. And it is not certain that the several types of prawn, some growing to sizes several times the others, respond to the same environmental conditions.

Since the subsequent stages of research require experimental facilities not available to the lab, the research results are pending extension work in cooperation with the government's Fisheries Department. With present knowledge it is assumed that a yield of twenty tons per hectare per year will be achieved initially, and that with more experimentation the yield could be raised to fifty tons. Assuming a net price of seven dollars per kilogram, a substantial income could be obtained from a ten-hectare farm that would provide food for local consumption as well as export. Some of the costs in productivity estimates were developed through a visit of a scientist to the United States and through talks with fish farm personnel in Unilever.

Location Criteria

Although the need for a development lab was seen some twenty-five years ago, it could not be established until there was an opportunity to obtain personnel with appropriate scientific training. The preferred backgrounds for professional personnel in India are engineering and science, but even so only one of sixty applicants from these fields will be chosen to join the company because of poor preparation and oversupply. Graduates with these backgrounds have demonstrated that they are more effective than those trained as pure scientists, who tend to be more academically oriented and less eager to get into industrial problems.

The main factor in deciding location, however, was that of

an already-existing manufacturing operation that needed technical support and that generated funds to support the research activities. Presently, the constraint on further research and development is the profit picture for HLL, not any criteria established by the Research Division in London. A secondary constraint is that the Indian government licenses all new product proposals, seeking to prevent resources from going into unnecessary product areas.

The government has, however, continued to encourage R&D through permitting 100 percent write-off of equipment in one year. It has considered permitting 150 percent expensing of R&D costs, but has not yet granted this incentive.

HLL has now to dilute its Unilever shareholding from 85 percent to 51 or 40 percent, depending on the final decision of the government under the Foreign Exchange Regulations Act. Against this background, it is useful to note that if HLL had been a joint venture with a dominant single partner when it established the R&D lab, it would have been much more difficult to create such a facility, relying as it did on Unilever support.

DEVELOPMENT LAB-BRAZIL

The Brazilian company, Gessy-Lever, has several labs related to its product lines. The company's five major product groups include detergents, toilet preparations, margarine and sundry foods, perfumery, and ice cream—out of which 310 products emerge. Each product area has its own manager with marketing, production, finance, and product development responsibilities; these report to a chairman, who in turn is responsible to the Overseas Committee. The centralized management at the national level covers financial, legal, and management services, physical distribution of goods, personnel policies, purchasing, acquisitions, and public and government relations. Despite this wide range of responsibilities, the staff at the central offices is small—for example, the "marketing information manager" has only two people working with him on overall marketing strategy, though he may draw on outside consultants or on divisional staffs. A General Manager's Committee composed of the heads of the five divisions plus the head of the distribution division and the financial and personnel directors meets each month. The general manager is responsible for R&D activities in his division. Until recently, there was no central lab, but some centralized R&D facilities are now being created.

History

The Brazilian company was established by Unilever in 1929, producing at first soap and perfumery. It added soap powders in 1952 and non soap detergents in 1959. In 1961 the company acquired a locally owned firm called Gessy, which was in powders, soaps, and toothpastes. Its sales were larger than the Lever Company's in Brazil, but there was no product development, and the third generation of the family was uninterested in running the business. This acquisition caused a spurt in the growth of the Brazilian operations.

In 1968 the company was divisionalized into soaps and detergents, toilet preparations, and margarine. Ice cream was added in 1973 with the acquisition of another Brazilian company called Gelato. Also in 1973, an industrial detergents company was acquired. And in 1976 sundry foods were begun on a small scale under the margarine division.

Each of the divisions has two basic activities—production and selling, with the latter the more important. For this purpose, the company has sought product improvement since World War II, and a gradual substitution for imports of materials and equipment. By 1972, all imports of materials and equipment amounted to less than 10 percent of the sales volume. By 1976 it was down to less than 5 percent. The largest import had been ingredients for the NSDs, but local petrochemical production gradually permitted substitution. Now even tripolyphosphates are bought locally. The most important import is packaging machinery, since high-speed machines for the packaging of toiletries and soaps are not available locally. In addition, some ingredients for flavors and scents (not perfumes) must be still imported.

This substitution of local materials has been a major responsibility of the development labs in each division. Since there was no central R&D for the Brazilian company, the budget for these activities was also not centralized, being met by each division.

The Chairman of Gessy-Lever observed that the status of R&D activities in his company reflected, in his view, the situation in the more advanced of the foreign-owned companies. That is, he had called ten of the international companies in São Paulo to invite the R&D director of each to meet with the head of the Research Division during one of his visits to Brazil, but he could find only one that had activities that could be called R&D. The consensus among other observers is that there are few affiliates of international companies that have yet progressed to a level of substantial R&D

activities in Brazil. Most of them remain at the developmental level or below.

Organization

There are nearly 5000 employees in the Brazilian company, over half of them in the detergents division; some 800 are in toilet preparations, 300 in margarines and sundry foods; nearly 500 in ice cream, about 125 in perfumery, and some 250 in the distribution division. Development labs are attached to only the first two of these divisions, with some 50 people in detergent development and 30 in the development lab on toilet preparations. There is a small unit of 5 people in margarines and another of equal size in frozen foods and sundries, with none yet attached to perfumery because the necessary work is done in the toilet preparations group. These lab personnel do not include those working in quality control and assurance.

The close ties of the development labs to manufacturing is seen in that organizationally, the product development manager reports to the technical manager in the detergent division, who is responsible for all production and product development in the division. Reporting to him also are the technical managers with responsibility for the two different plants (São Paulo and Valinhos), who in turn have a quality control group as well as one concerned with process and packaging development alongside their engineering and production departments. At São Paulo some forty people are in quality control, with sixty at Valinhos; responsibilities in this group include test procedures, quality control, and analysis of specifications. In addition, the process and packaging development group, about one-third the size of the quality control group, provides technical service for in-plant processes and packaging procedures.

The product development manager, who has about fifty people in his group, is responsible for development of product formulas related to fabric washing, toilet soaps, and dishwashing liquids, and for consumer research, product evaluation, and technical information. A small group will eventually be added to analyze specifications, drawing some of its members from the existing quality control groups now doing this type of work (see Figure B1.1).

It should be noted that the quality control unit does not report to the factory manager, but to the technical manager at the site, remaining separate from the product development group. Any production from the plant which falls outside the rigid specifications applied by Unilever internationally must be approved for sale by the

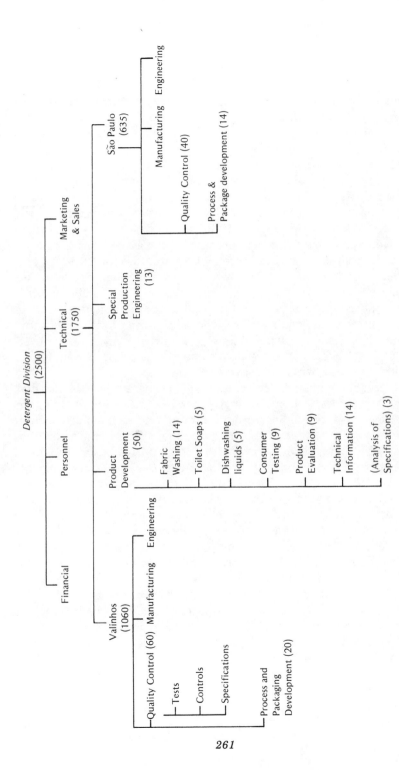

Figure B1.1. Organization of Development Lab-Brazil.

marketing manager for that particular product division. The process and packaging development unit is linked to problems of production at the factory site and obtains its assistance, when needed, from the product coordination units in London—and indirectly from the central labs or Research Division.

Some new quality control activities were introduced with the acquisition of Gessy in 1961, and by 1972 routine quality control operations were separated from the production unit, with all production judged against the international standards of Unilever. To perform its functions adequately, a quality control lab requires analytical equipment to test whether the products are maintained under the specifications required. Since 1961, quality control facilities have expanded three times, along with the expansion of product range and production volume. Product volume has increased 12 percent per year for the past four years, with accompanying increases in quality control activities. The skills required in quality control are not those that are needed for product development, since the latter requires a knowledge of basic science. When individuals show an interest in applying basic scientific knowledge, they are moved into the development group, where some will be put in the analytical specifications unit, which is yet to be formed.

The organizational structure in toilet preparations is somewhat different since the technical manager has two managers reporting to him who are responsible for production on the one hand and technical services on the other. The production manager also has a unit concerned with plant and process development, while the technical services manager has not only product development but also packaging and quality control. This location of the quality control activity reflects Unilever's concern that it not report directly to production or factory operations but to R&D and marketing units.

The professional in each of the product development units responsible for a given product line is a university graduate and will have one or two technical assistants and one or two helpers to carry out the program activities. All of these individuals are Brazilians; in the toilet preparations division, only four managers are expatriates, and the company is presently in the process of recruiting and training replacements for three of them. The most difficult one to find has been the replacement for the head of the product development lab.

Since the educational system is modeled after the French, training even in the engineering programs is largely theoretical, so that engineering graduates have to be trained on the job for industrial activities. This is done by bringing European engineers to Brazil or by sending locals to universities for additional courses. Each engineer has a personal development profile and is offered a sequence of courses within the company itself. Occasionally the engineer is sent back to the university for a Master's degree, though chemical engineering in the Brazilian university system is not yet adequate. There are very few Ph.D.'s available, since they tend to stay in teaching, and the company is not yet doing the type of research that would be interesting to a Ph.D.—though they anticipate doing so as they proceed with their reorganization of R&D activities.

Projects

The research program of the two major development labs within Gessy-Lever are oriented primarily to adapatation of products developed initially by Unilever, since each of them requires some translation both as to processes and to ingredients. Even in toilet preparations, where the Unilever Coordination is attempting to establish a set of international ingredients and formulas, it is not expected that they will fit perfectly with what is available in Brazil, and adaptations will be required.

The detergent division has asked to be kept informed by Port Sunlight, where there is a section devoted to dealing with overseas product needs. For example, the division is presently engaged in discussions with the lab in Germany concerning ingredients of a new product that is being developed there. The Lever Bros. company in the United Kingdom might help on a specific translation needed by the Brazilian company, and it is to the Lever Bros. development lab that Brazil would turn for this kind of help. Each translation effort by the Brazilian company leans heavily on these facilities in the United Kingdom.

Similarly, in toilet preparations, the deodorizing ingredients in a toilet soap were not compatible with some of the perfumes available in Brazil, requiring some comparative testing on different perfumes; some of this was done at Isleworth. In another instance, the lab had to scale down the super-fatty ingredients in toilet soap because of the high cost of the ingredients in Brazil. It was successful in reforming the soap to cut the cost without a reduction in performance; the product is now being adapted for European use. Differences in the availability of certain ingredients are so significant that some research in Europe is simply not applicable in Brazil. For example, Europe is now insisting on biodegradable ingredients, but only nonbiodegradable materials are available in Brazil. To determine how to use these kinds of ingredients, the lab drew on earlier research and older products in Europe, though even these had to be adapted to different inputs.

In one project, the detergent lab developed a powder with new ingredients (not Brazilian) and then had to undertake the lengthy task of convincing Port Sunlight and Coordination of the value of using the ingredient for the purpose of cost reduction.

Some of the development lab's program is related to import substitution being done by suppliers. Technical assistance is provided on specifications for chemicals (tripolyphosphate, for example), with Port Sunlight testing some of the samples and Colworth testing for toxicity. Once, to get non-ionic detergents for dishwashing, the lab helped the Atlas affiliate of ICI on specifications, based on a good deal of background work with the Colworth lab. In fact, ICI and Colworth were working together on the chemical in Britain.

In its project on hair research, the lab's ongoing task is to reduce import costs of ingredients—both to substitute local ingredients and to cut production costs. In this it works with the suppliers (usually affiliates of other foreign companies) on changes in specifications of materials. These local materials may be more costly than imports, but restrictions on foreign exchange prevent getting licenses when local supplies exist—though the latter may not be of the same quality. These efforts have at times cut costs, averaging out to about 60 percent in a number of materials. Over the longer term, the Director of the lab is attempting to get professionals to think of new product forms; professional scientists have the ability to "blue-sky" a bit and to come up with some new formulations. However, they must and frequently do recognize that their efforts would not necessarily be rewarded by adoption by marketing.

In the lotions unit, the lab has done all the compounding of materials, working mostly on the problems of the degrading of perfumes when different colorants are used. In skin research, the major emphasis is on deodorants and on the effectiveness of different applicators. Some development work is also done on talcum powders and shaving creams as well as skin creams.

Consumer research is very much needed in the TP area especially, but it is also quite expensive. Little research exists on Brazilian consumer habits regarding the specific use of products in the home. There is also relatively little hard information about the use patterns of competitors' products, or about product preferences. The lab is moving toward consumer research, having received approval to do some in its product development projects. They will set up a thousand test users in one city and supply a variety of products; these users will reply to a set of questions concerning performance. This project will require a professional scientist, a statistician, three or four helpers, and a car—all at a cost of 200,000 cruzeiros (Cr. $) per year plus the cost of the products used. The total cost of the tests will still be only equal to about what *one* such test would cost if done by a consultant.

Most of the activity in the area of packaging research centers on finding local materials, taking off from U.K. specifications. Adaptations are required to local machines and to the competition—trying to match as closely as possible volumes and sizes used by others so as not to lose in the market. Available local machinery may be less efficient than is required by Unilever cost and performance specifications. Mixing equipment can be bought locally, but some fill-in lines must be bought abroad. Differences in packaging and product form are important in the Brazilian market; the required adaptations are different for differing geographic sectors and market segments, but the company does not know enough about each of these as yet.

The labs will need to do much more consumer research than they have done in the past; e.g., the toilet preparations line of Unilever needs to be extended to meet the current demands of the Brazilian market. One of the lab's responsibilities is therefore to constantly survey the country's toilet preparations market, regardless of what exists in the Unilever line. Neither the marketing group nor the R&D group knows how to do consumer research; most of what is done in this field at present is to get a quick payout with little cost, such as interviews at a shopping center in São-Paulo. This obviously is not a good sample, and is useless in other parts of the country. (The differences between Paulistas and Cariocas [Rio] are substantial.) The lab is doing some attitude research, but it is not deep enough to be useful for product development purposes. Still, the price of success in this effort is so great that the company is reluctant to commit substantial funds for consumer research on cosmetics and dental products. The cost of getting a diversified sample—geographically and racially—in such a large country is staggering compared to the potential payout to the company. This lack of research in the field presents a serious obstacle to Isleworth's setting up an R&D center for product development for LDC markets; such a center simply would not have adequate consumer research information at present.

A final program for the labs relates to regulatory affairs. As indicated earlier, there is virtually 100 percent reliance on U.K. labs—Colworth tests every product before released and the head of research division personally approves any adaptation by overseas labs. The company makes no effort to correlate its standards with the Brazilian government's regulations; it simply seeks to get Unilever's standards accepted. There is therefore no communication on the formation of standards for the Brazilian economy as a whole, simply a negotiation with officials approving new products. Still, the increasing injection of governmental interest in import substitution and restraints on imported equipment and pricing has caused several officials to spend a large amount of time on government relations. The head of product development in the detergents group spent nearly 90 percent of his time in the last half of 1976 on government relations; the toilet preparations division also has an expert scientist half-time on government relations. Both recognize that corporate safety testing will eventually be required in Brazil to improve the chance of approval of new products.

Coordination with Central Research

Few of these projects could be carried out without close liaison with Central Research in Europe. The Isleworth lab now has an overseas unit specifically dealing with the needs of LDCs, and is examining the problems of product adaptation for taste and the necessity to use local materials for both cost reduction and foreign exchange savings.

The procedure by which Brazil finds out what is going on in Europe is the same as India's—the Technical Information Service— but the flow has expanded since the head of the product development lab came from Britain, simply because he has been able to assess the information from a scientific and commercial viewpoint. This official takes trips twice a year to Port Sunlight to assess detergent developments and finds out what they may have rejected that is of interest to Brazil simply by asking around. Other lab officials also visit European facilities; these visits to the United Kingdom extend also into the process and packaging development group.

Since the development lab in detergents is largely concerned with existing products and processes, technical assistance comes directly from detergents coordination in London—not from Central Research, though Coordination may get some assistance from Port Sunlight. If a research project is itself involved, the Brazilian group would go through the Research Planning Group in Research Division. A current major cooperative project with Port Sunlight is the improvement or introduction of new products in fabric washing powders. Brazil is providing consumer research and evaluation, under the direction of the PS fabrics-washing development manager, who is spending 80 percent of his time on this project, developing studies on washing habits in Brazil. He has an assistant scientist and three technical assistants. The chairman of the team is the Brazilian detergents marketing manager; others include the Brazilian head of Product Development in the division, a professional from PS, one from detergent Coordination, and two others from the marketing and technical areas of Coordination.

Port Sunlight is studying the physical chemistry of Omo (the Brazilian washing powder) in local conditions—cold-water washing, for example, and very soft water. Unilever Research Division sent two individuals down to examine consumer perception of their needs, relying on methods of analysis developed in Port Sunlight. They invented a new approach to simulate washing by card-sorting (the number of shirts in the types of water and the detergents) so that computer analysis could be used to divide and quantify the different washing procedures. A representative sample of procedures was identified for testing products in order to determine the physical chemistry. Customers are questioned to learn the problems they have in areas such as stain removal. Port Sunlight experience is being applied to these problems by analysis of physical chemistry. PS is also seeking the best bleaching results with oxygen bleach, without concern at present to the cost factor in materials used in production.

Port Sunlight professionals and the Brazilians will jointly analyze the results. Communication with the Brazil development lab has improved greatly with the assignment of a PS professional as its head. Although he is only one man among ninety-five, his knowledge of PS procedures and personnel makes an exchange of information much easier.

In order to undertake its part of the project, the Brazilian company had to place a scientist in charge who would be able to interpret the experimental results. Port Sunlight sent a detergent technologist to help train Brazilian personnel. The technical assistance employees are not up to the level required and need additional training. Eventually a physical chemist will need to be trained, one who has particularly good communication arts. The Brazilians have wide-ranging capabilities, some with degrees in advertising, but personnel in technical spots are sometimes found to be without formal technical training. Top management is therefore required to do a great deal of technical training in the R&D area, though they do not always have time to do it adequately. One Brazilian was sent to Port Sunlight for two months, and returned to Brazil only to leave for another company at twice his former salary.

One additional problem in the joint project in Brazil is that the government has made it difficult to import test products or equipment. The company therefore has to use its full production scale equipment to produce merely 100 kilograms of product.

The Brazilian detergents development lab has yet to engage in a "future product" project, but it does anticipate such a project, possibly employing technology in non-European countries, or something entirely new. A joint project exists with Research Division that examines long-range R&D development, with frequent meetings of officials in Europe or Brazil. Much research is needed on washing practices and materials in Brazil, since both are still poorly developed.

Cost/Benefit

The benefits of the lab's activities are shown in the reduction of cost of production. Over the last two years the record has shown a cost reduction greater than the cost of product development and technical services within the company.

The reductions in imports are not simply cost savings, but will also permit the company to continue to produce, given the increasing constraints imposed by the government on imported materials and equipment.

Reorganization

The Brazilian company has reached the point where it needs to consider taking the next step up the R&D ladder.

The first step in the reorientation will be to bring together scientific and technical people from all divisions into a single lab, adding a research capability to the present development orientation. Product development labs would still report to the product divisions, but a central research capability would be opened up. This is seen as necessary in order to keep the interest of some of the more capable researchers and also to attract some with Ph.D. degrees from the universities. With this added capability, there might be a possibility of shifting some of Port Sunlight's work for developing countries into Brazil.

The first unification will be in the detergents division, but others would be eventually brought in as well, though they would be added at subsequent stages as more room is made available in central headquarters buildings. These next steps may be four to five years away. It will take this long for the lab facilities to become comparable to those in India, by which time R&D facilities might be moved to a completely separate site, as they are in Anderi.

These contemplated moves do not appear to be upsetting those presently working in development labs, since many of them do not understand what applied research is, nor do they see it as disturbing their current work in translating U.K. developments. No disturbances are in fact likely, and a central R&D capability will be developed merely by bringing those who are able to do applied research into a single unit.

Community Relations

In order to build up a better resource for future recruitment of scientific personnel, the company is attempting to assist in an exchange of scientists between British universities and São Paulo State University and that in Campinas. Two years ago officials talked with the Brazilian Association of Science and the National Research Council of Brazil, identifying several universities that might profit from such exchanges. The two at São Paulo and Campinas were considered to have sufficiently advanced curricula to benefit from assistance, and both are close to company plants.

The company also helped set up a Colloid Science Group (relating to research in soaps, detergents, toiletries, ice cream, and margarine) at São Paulo State University as a subfaculty and a scientific specialty. Port Sunlight officials are giving assistance on a consulting basis, and will send a colloid (bench) scientist to work for six months with the faculty at São Paulo training the first group of entering students. The company has also sent the head of the Colloid Group at São Paulo State to Port Sunlight for three months for an exposure to the field and to learn who is doing what in it.

In addition, the company has given two scholarships for studies leading to the Ph.D. degree in Brazil and two for Ph.D. studies in the United Kingdom at universities or the Port Sunlight lab for postdoctoral work. All four of these professionals will eventually teach in the Colloid Group at São Paulo State and be available for consultation with Gessy-Lever.

The British Royal Society and the Brazilian National Research Council have been given grants with no strings attached for the exchange of students or professors between Latin America and Britain. It is up to them how they use the exchanges, informing the company merely of the expenses to be reimbursed. There is no link to research interests of Unilever in this funding.

The objective of all this collaboration is to achieve greater expertise within the universities in chemistry and chemical engineering, thus raising the possibility of better recruitment. The company is expecting closer collaboration with the university at Campinas in organic chemistry, but this has not yet been developed.

At present there is very little academic research in the universities, though what research is done is academically oriented, attempting to fill in gaps in the literature. A slow process of redirection of academic and engineering research towards industrial needs is occurring through the Brazilian National Research Council. This lack of interest within the universities is also matched by a low level of industrial research within Brazilian companies—even within affiliates of international companies.

Consequently, there are very few consultation arrangements between the company and professors at the universities; strong ties exist with only a few faculty members. Similarly, there is little contact between the company labs and governmental laboratories, though there are two professionals who work with government labs at times.

The absence of appropriate academic resources remains a bottleneck to the enhancement of R&D activities within the company, though the company is taking steps to make certain that the bottleneck is not too tight when it becomes ready to upgrade its activities. For instance, some university labs are used for special tests, and these contacts are likely to increase in the future.

Governmental Relations

The Brazilian government provides no special incentives for R&D development; however, the Brazilian National Research Council has helped Gessy-Lever with its relationships with universities, since it does see a need to develop ties between universities and industry. (The U.S. National Research Council has had cooperative projects with its Brazilian counterpart to stimulate this interest.)

The Brazilian government's increasing interest in environmental protection has elicited visits from both detergent and toilet preparations Coordinations in Unilever, as professionals have come to talk with the government on the production processes and their effects. It is anticipated that the company in Brazil will have to develop its own expertise in this area in the future.

Looking ahead, the company anticipates that local participation may eventually be required, but not within the next five to seven years. The capital market is not strong enough to give a fair price for the shares, and it would require considerable effort to turn up sufficient investors to buy a significant percentage of the company. Nor would it be easy to find a joint-venture partner who could make any kind of contribution. A joint venture, in the minds of the officials in Brazil, requires a synergy which produces five out of two plus two. A joint-venture partner is unlikely to understand the need for R&D expenditures, and even local shareholders are likely to consider such costs as necessarily reducing their dividends. Local partners, however, would add no technology or management, and only a merger would provide a factory or a market position. Most potential partners are seeking to sell out rather than to merge, but the government is seeking to keep them in through low-cost loans from the National Development Bank.

But Unilever, as many other TNCs, prefers to retain at least majority ownership. Without the majority control needed to select management orientations, the company fears that constraints would be imposed on the R&D program, simply because U.K. labs would not want to give up readily what they have spent large sums of money to develop. A majority local partner might seek to pull away from the Unilever line, forcing Unilever Coordination to withhold product developments or assistance in R&D.

However, it does seek managers from among Brazilian nationals and has raised their number rapidly. The number of managers has increased from 85 in 1970 to 400 at present. Only 36 of the 400 throughout all five divisions are expatriates, and these have an average of only two and a half years in Brazil, since they are mostly on loan with the requirement to train their replacement.

Future R&D in Brazil

One possibility of a new orientation in R&D in Brazil was discussed with several officials—that of developing a center for R&D in Brazil for all of Latin America. It was agreed that a single center could be set up in Brazil on hair or skin, simply because the country includes a wide range of ethnic variations. A particular obstacle is the unavailability of local scientists, who would need postgraduate work or experience in skin properties and in research methodologies and safety testing; local expertise needed in chemistry, biology, analytical chemistry, and consumer research simply does not exist. (For example, it required three months of recruitment to find a woman who could take a managerial position in skin cosmetics, and the company had to settle for someone who had no experience in skin treatment itself.) Scientists could conceivably be brought from Europe, though they would be difficult to recruit, and Brazilian government approval for such transfers is unlikely. A second obstacle is the cost of such a lab—estimated at between £1.5 to £2 million per year—which would require £100 million of sales to support it.

A third obstacle would be that other countries in Latin America would not likely welcome the research coming out of Brazil. Nor might other developing countries, since there is a feeling that such research would not be at as high a level or as effectively done as at one of the major European labs.

Finally, a Brazilian center would also run into the difficulty that other affiliates throughout the developing countries would not yet be technically able to adopt or use its results. It will therefore take some years before an international company can achieve such an integration of R&D activities among affiliates operating in several developing countries.

FUTURE R&D IN DEVELOPING COUNTRIES

Unilever officials in various divisions of the company were asked to assess the future of research activities in the developing countries, taking into consideration the stages through which the development might occur and the obstacles that exist. Given the heavy orientation of the company toward consumer needs, it would appear that substantial research would be required in a number of countries where the market might be expanding rapidly. As might be expected, there were divergent views as to the possibilities and desirability of Unilever undertaking significant R&D operations in the overseas countries. These differences showed up in a discussion of the objectives for such research, the relationship to the Research Division, the scope of research in the developing countries, and the geographic location of labs.

Objectives

Two major objectives appear to be important in the minds of Unilever officials: one is the need to do research on indigenous inputs of materials in a variety of product lines (for example, India and Brazil), and the other is a response to political pressures that the company provide a scientific presence in the host country. In connection with the first, Unilever has recently realized the need to do more research on foods, though this move is still in the inception stages. It also recognizes that the markets for detergents are increasing quickly enough in LDCs to require some local research in that product line. Therefore, the greater the demand for products that reflect indigenous needs and tastes, the greater the need for consumer research and product development activities. As the head of the Vlaardingen lab stated in a speech in 1975, "As consumers [in the LDCs] start to demand an increasingly varied range of products, the requirements of greater technological sophistication will also have to be met. It will probably be easier to achieve this if the finishing touch in product development can be applied more on the spot. Working at a distance is in such a case no longer efficient."[8]

The objectives of Unilever R&D expansion in developing countries are to gain new and improved products in LDCs and to promote better government relations (including those with regulatory bodies) and improved relationships with opinion-forming groups in the host countries. The company expects to establish R&D units where they can be commercially effective and also help host countries meet their long-term goals. The company anticipates that those who do not respond appropriately in the R&D area will lose permission to import technology from the center.

The Unilever Board has decided that a greater priority must be given to research by the overseas companies, considering that—if nothing else—less research is done there per pound of profit than in Europe.

Besides responding appropriately to government concerns, Unilever officials recognize that they need to explain what the company is doing in research, so that governments understand what they are *already* doing. They see a possible further move to higher-level technologies in the production processes, in order to provide more advanced experience and training for local workers. Some techniques of production are already being developed outside of Europe that might be applicable to the developing countries, such as the oil flotation process employed in South Africa.

Having said this, Unilever still expects to examine carefully potential locations of any new labs, since it is no advantage to establish a non-useful or inefficient lab. The company expects that many would be local development labs tied to local manufacturing companies. However, as they proceed, some could expect to become "affiliated" labs and a few eventually part of the Central Research Division.

Relation to Research Division

Despite the fact that local research activities in developing countries are likely to be tied closely to the producing facilities in each company, the Research Division will still have a central role to play in the expansion of these activities.

The Research Division has always responded to the needs of the Overseas Committee, but the supporting research has generally been carried out in the European labs. Each operating company in the overseas group can pick up the results that are applicable to its markets. The Research Division is looking at a couple of countries in each region to examine their needs, but the decentralized structure of Unilever does not lend itself to a regional approach to production and distribution, though it might enlarge the market and support further research. Overseas companies are tied individually to the European labs, where all overseas research is synthesized. (Of course, research on some foods would not lend itself to a regional approach—for example, that on babasu in Brazil. Nor can a regional toothpaste be developed for Southeast Asia because of differences in packaging and flavoring among the various countries.)

Some officials see a potential conflict between the objectives of the manufacturing companies, which will want locally oriented research for their commercial and political survival, and the Research Divisions, which will want a centralized and coordinated program to reduce costs and achieve efficiency. For example, research on seeds will lead to research on oils that can be used someplace other than where they are produced, eventually in products that may be exchanged across national boundaries. Thus, the design of the research has to look at consumer needs in several countries, which can frequently be done better at a centralized location.

To some Unilever officials, it appears quite difficult to set up a specialized research unit abroad, unless it was oriented to developing indigenous inputs for an indigenous product. For example, the project on reformed meat would lose some quite significant inputs from Colworth if it were cut off and located in Brazil. There it would not have Colworth's muscle analysis, which is necessary to determine the best ways of cutting and reforming the meat. It would not have capabilities such as electronmicroscopy, and it would be far removed from the major markets in which the meat would probably be sold. However, a project on fish farming could be put in India without a great cost to the Colworth program. It would be closer to the market in LDCs and not as reliant on sophisticated equipment or technology. Even so, India would probably not have all the necessary capabilities. Some officials have commented that Unilever "can't do anything small."

Unilever is not used to running dissociated labs—that is, any lab not tied to an operating company—save in Europe. However, the closer to local operations and the farther geographically from the Research Division, the less coordination and potential efficiency exist. But corporate headquarters can never be divorced from any affiliate's acts or efforts, since the company's total reputation is at stake; therefore a pull arises to maintain central control or at least surveillance of research in the overseas countries or to tie small, local development labs to an overseas RDAU.

It was estimated by some officials that future research programs in detergents in LDCs would be organized to centralize research on common problems or to specialize in LDCs on parts of research—for example, washing at different temperatures. But one of the obstacles to a spinoff of projects into developing countries is the difficulty of developing an appreciation at the center of the differences among the developing societies' product needs and consumer preferences. Unilever has been putting more effort here in the last few years, but it is still often surprised by the way consumers handle foods and detergents. Another obstacle is the need to have someone at the other end who has a similar level of professional expertise and is putting the same intensity of effort into getting the results of research applied—that is, someone at the other end needs to pull while someone at the center pushes.

The Central Research labs have contributed directly to technical expertise in developing countries through a variety of programs. Programs undertaken through the Food and Drinks Coordination provided some agricultural advisory work, some technical aid, and cooperation with some international organizations working with the developing countries. Within the agricultural advisory work, there were six programs in different countries on growing sunflowers and for testing several varieties in the various countries.

The sunflower research was on selection, growing, effects of different climates, drooping heads, and the effects of different soils. This research has been given to governmental institutes in Mexico, Turkey, and other countries. It literally saved a Turkish crop that had been threatened by disease. The Unilever interest here has been to expand the sources of sunflower oil on the open market, and it is still doing research on a dozen varieties. This research is part of a larger program in horticulture on vegetables of different varieties, which in turn requires work on soil cultures. This calls for a team of agri-horticultural experts, chemists, and analytical chemists with wide-ranging capacities.

The Vlaardingen lab has the very best experts moving around the world doing tests on sunflower growing. The results have been made public for all interested, to ensure that crops will be larger and better. R&D officials have been able to commit resources to this project despite its not directly yielding a product for Unilever, because it does fit within a wider product-oriented program.

This research has been expanded into rape seed, seeking some disease-resistant strains and some with a reduced percentage of undesirable acids. After much research, the labs have been able to produce a strain without the undesired acid composition by actually cutting the seed to eliminate the part containing the acid. The research covered the responses of the seeds to different soils, climate, the phases of planting, characteristics of the seed, and processing problems and extraction. This kind of work could not yet be done in the smaller labs available in the developing countries. They would have to rely on the capabilities of the central lab, for the rape seed program required ten teams and was itself built on a much wider nutritional program.

Advisory programs also existed in the early 1970s for palm oil development in Ghana, on vegetable growing and canning and fruit preserves in Turkey, on mass feeding programs in India, on vegetable growing for frozen food exports in Morocco, and on agricultural support for industrial development in the Mekong Delta. Technological assistance stemming from research in the Central labs included the use of vegetable protein in institutional food in Iran, extruded vegetable protein as a rice substitute in Indonesia, soya milk processing in Sri Lanka, oil refining in Liberia, integrated meat development in Kenya and Botswana, fishery development in Venezuela and Peru, and protein fruit development in Nigeria. These last four were in cooperation with the Food and Agriculture Organization (FAO).

Agricultural programs in the LDCs are dominated by the FAO, the Agency for International Development (AID), and United Nations Development Program (UNDP); consequently, Unilever has restricted itself to rather small-scale programs. However, it has been working on an air-inflated greenhouse to help diversify agricultural output, the type of program that has been welcomed by LDC governments. Where it has an interest in large-scale programs, it has had to move slowly and through governments, since commercially-oriented agriculture would alter a country's culture.

The establishment of an R&D Application Unit for Latin America and other regions will be influenced by the size of the market, similarity of markets, and the ability to be supported by a central R&D lab. Such an RDAU could not be put in a development lab within a Brazilian company and at the same time serve all the other companies in Brazil—there would be too much jealousy and suspicion to permit this to be successful. It has to be separate from the operating companies to not seem partial to any one company in the system. (Unilever did do development work for Scandinavia in Sweden, and the other countries distrusted it.) Without the backing of a Central lab, an RDAU for Latin America would not have the sophisticated service that would be needed to make it effective.

One lab official argued that an RDAU for developing countries could be established at Colworth, oriented basically to problems of food processing; but even this would leave some difficulties unresolved because of the distance from the operating companies. Mere physical distance will make it more difficult to handle than the European RDAU, but RPG chairmen are among them now taking as many as forty trips a year to overseas companies, and the younger professionals coming up are also willing to exert themselves in this way. The company would send some of its best men, as it has in Brazil, to help establish these development labs; it would then expect to lose them to the operating companies—at least for a while.

Development of an R&D presence is illustrated by Unilever's activities in Nigeria. The company was able to help the government find food substitutes, easing its nutritional problems, but without direct profit to the company. It was also permitted to produce toilet preparations, which were profitable. Now, however, it needs a small lab of four to ten people to keep production going, which, once established, will provide employment for scientific and technical personnel. These workers will be involved in quality control and some research on mineral oil substitutes. Later, they will become involved in problems of different types of hair and hair styles. These are not the same even throughout Nigeria, reflecting different customs, but the company will still have to produce a single (undiversified) line of products.

LDC Research Programs

The first program stage for an LDC lab is the establishment of quality control and quality assurance related to manufacturing. Gradually, depending on the market and the availability of manpower, product development can be begun. There is, however, no necessary functional relation between quality-control responsibilities and functions of a development lab. Very small operations in food research can be begun simply by one professional in a kitchen, mixing powders or dehydrated soups or sauces or flavorings. However, if this small operation is tied into a larger system, quick answers can be provided to new questions, preventing redesign of the wheel. Such exchange of information requires that the man in the kitchen be a sophisticated professional, so that he can ask the right questions and understand the answers.[9] He also needs to know how to adapt the answers to the local market, meaning he must pay attention to local market needs and government objectives.

Given their separation from Central labs, labs in the developing countries need to be able to make quick responses on their own; the problems they face change as governments shift their economic and social welfare objectives. Governmental requirements concerning new types of products, local content, and safety also alter definition of the research problems. This makes the establishment of effective programs in developing countries' labs extremely difficult, because most R&D labs are "programmed" for yesterday's problems, and top management for day-before-yesterday's problems. Present R&D managers have a difficult time envisioning the future, with its changing needs in developing countries.

Specialized labs in LDCs would be very difficult to orient and control. A specialized lab on "cloves in Zanzibar" would be difficult to staff and operate because it would need an understanding of markets (which are outside the country) and a sophisticated expertise in problems of growing agricultural products. It would also need a good, efficient, scientific manager and dedicated personnel. This is not to say that it cannot be done, just that it is most difficult.

The product development effort in Indonesia is oriented towards use of local materials and nutrition problems. Java is growing soybeans for personal consumption only; there is no commercial farming. This shift will probably be made on Sumatra so as not to disrupt Javan customs. In order to accelerate the necessary shift, research programs would be needed on soils, processing, distribution, and use of soybeans, probably in cooperation with the UNDP and FAO. Many Javan university research teams are working on soybeans, but are not significantly oriented toward commercial uses; Unilever could assist in reorienting the research. Any effort it would make should be tied to the operating plant in Indonesia for its support and for greater realism in the research.

To get Indonesian support for the soybean research, it might also send experts to talk with officials on quite different projects in which the government itself is interested, such as arteriosclerosis, which is rising in wealthy groups in Djakarta and is found even among Ministerial officials. Unilever also has a research program on growing palms to produce oil, aimed at improving plantation efficiency, based on an analysis of vegetable tissue. It is giving the results to the Indonesian government, and still has a long-term program on the subject.

The recently established lab in Turkey suggests the scope of a new facility in an LDC. The objective of this lab is fourfold:

To extract and submit for public benefit the country's unused resources;

To establish links between research institutions in universities, industry, and government;

To establish a laboratory for research on the growth of oil seeds;

To increase local contributions to research.

The projects initially undertaken included assessment of cooking procedures and industrial resources, utilization of forestry products, assessment of perfume plants, and synthesis of perfume components. Several studies were undertaken with reference to oil seeds and nuts—sunflower, rape, groundnut, tobacco, tea, grapes, and pistachio. The trials with the new sunflower seeds are in cooperation with university faculties, agricultural research institutes, and the sugar company of Turkey. The prime objective in the research on rape-seed oil is the removal of erucic acid for the growing of rape seed in unused areas of the eastern part of Turkey. The demand for groundnuts (peanuts) has increased for snack foods, and research is being expanded to increase the acreage planted in groundnuts. Tobacco seeds are normally wasted by the peasants in Turkey, but they yield an oil useful in cooking and in industry, and the cakes can be used for animal feed. Tea seed is also a source of cooking oil, yet it has not previously been used this way in Turkey. Grape seeds, which yield about 13 percent oil, are now simply thrown away in the manufacture of wine and other products. Pistachos grow wild in the country, and their use for oil is presently very small; the research is aimed at expanding the potentials of this nut.

The Turkish program also includes research on the many scented plants grown in the country, particularly around the Aegean and the Mediterranean regions. It will also examine the availability of raw materials from which synthetic perfumes can be made.

These are the initial components of the research program. All are directed at increasing available supplies of products needed in the country, based on indigenous materials.

Location Criteria

Even after it has been determined that research shall be undertaken in developing countries, it cannot be located in every one of the countries, nor to the same extent in each of the labs. The criteria for location will be related to the existence of manufacturing operations, the availability of support facilities and easy communication, and governmental restrictions.

Operating Company. The primary locational criterion would be the existence of an operating company in a given country, since there is a need for local administration and a local source of funds to take care of the services required by the personnel in R&D. An isolated lab requires too many administrative and support services to be economical. Also, there needs to be a visible application of what is being done and an urgency to it, or else the lab becomes academic in its orientation. The structure and product line of the company would dictate the type of research undertaken by the lab, its eventual size to be determined by the size of the markets served by the operating company. This market does not have to be merely national; it can include exports.

Facilities Available. Another criterion is that related to the facilities available to the lab. Not only must there be administrative support, but adequate communication links back to the center are required. If they do not exist, the personnel at the center—who are responsible for success of the overseas effort—cannot follow through. In addition, there is a need for a pool of scientific personnel—not just Ph.D.'s but also a tradition of R&D activity, even if this has been gained abroad.

Further, scientific work requires physical facilities that do not add to the frustration of the scientist. This means good sanitation, good housekeeping, adequate supply of inputs, and prompt mail and telephone communication, not only locally but internationally. The need for such facilities in a development lab are in a sense greater, or at least different, from those of a Central lab. For example, a small lab needs greater access to outside data banks than do larger labs, who can collect their own data inputs. Access to central data banks is difficult for many developing countries because it has to be done electronically and by satellite. Facilities for this simply do not exist in many countries, but if these various facilities are not supplied appropriately, the cost and frustration become so great as to make it advantageous to shift the operation back to the center.

Although training is obviously necessary for the scientists at such a lab, there is a danger in training them very highly. Training of development personnel beyond the first university degree gives them an image of themselves which prevents them from working on the factory floor, where it is necessary to translate research results into production. Training beyond the necessary levels is also costly; chemistry, at lower levels, is a simple mixing of elements and ingredients, once it is known what to mix. If the formula is provided from the center, no really advanced knowledge is required to do the mixing at the lab.

Finally, a higher priority must be given to support from governmental institutes, which can provide substantial basic background research; governmental support for scientific education is necessary. For example, Turkey has had a long history of education and science based on a close relationship to the German system and relying on German personnel. Consequently, it has a good nucleus of technicians around which to build developmental work.

Governmental Requirements. A proliferation of governmental requirements that R&D activities be set up would pose serious problems for Unilever (or any major international company). The company simply could not put R&D activities in every country where it had a manufacturing operation. However, *some* scientific and technical work will be required at each location simply for the purposes of technical services and quality control.

If the company is pushed beyond this stage by government regulations, it could cut off a few pieces of research in a country such as Brazil in order to achieve higher levels of scientific research, hiring fewer than ten professionals on a quite specialized activity—but, as discussed above, this also creates some problems. The selection of a country in which to set up a specialized lab would depend on the leverage that such a lab would have not only on governmental relations but on the future commercial development of the company and its future role in Unilever research.

One of the more difficult problems in determining location of R&D activities in the future will be the government's attitude toward joint ventures. However, from Unilever's viewpoint each case will stand on its own. There is a small joint venture in Portugal in which the partner wants all the R&D assistance he can obtain. Assistance was initially offered free by Unilever, focusing on methodology, problem identification, and research design. Once the partner saw the usefulness of these activities, the venture began to create its own development lab at its own expense.

The willingness to provide extensive scientific and technological support overseas depends on the type of partner and his ability to use the information outside of the partnership. In India, for example, the partners are private individuals, holding shares sold publicly, and though the ownership by Unilever will go down to 60 percent, the Overseas Committee of Unilever sees no significant effect on the R&D program.

The alternative of separating the R&D lab from the manufacturing company in order to retain 100 percent ownership is not attractive to Unilever, since it would divorce the lab's activities from the production needs of the manufacturing company. A related problem would arise in the ownership of patent rights and the charges to be made for the assistance given the lab in the locally-owned joint venture. The scope and nature of cooperation would be changed fundamentally, but Unilever has not yet prepared a policy for this contingency.

Summary

Summarizing the many conversations with Unilever officials, the following maxims for creation of effective development labs in LDCs can be set forth:

1. The establishment of an R&D lab in a developing country requires a large enough business locally to support a lab of some thirty scientists, ranging over several products.
2. Research problems should be sufficiently peculiar to the country so that they can be more effectively tackled through the local lab than at the center.
3. A national science community—such as university institutes, governmental institutes, or numerous R&D labs in other companies—should exist to support the local unit.
4. An adequate number of qualified scientists should be available.
5. R&D labs should not be established unless there is an *economic* basis for doing so and a *commercial* benefit to the company— never simply for public relations. Without a related commercial market, it is difficult to determine how large to make a lab. How should one be set up that is specialized and unrelated to the commercial operations in a country that is simply insisting on R&D? On the other hand, there are obvious advantages to setting up a local lab to develop indigenous products for a growing market and to help in their processing. Good public and governmental relations would come out of sound work by the lab, but it should not be set up solely for this purpose, for it could never demonstrate to government officials its usefulness.
6. Management of the operating business should have an understanding of R&D and a belief in its long-term results. R&D

has to be conducted even in lean years to provide proper phasing between R&D output and commercialization of improved or new products.

7. Other parts of the operating business should be sufficiently sophisticated so that the results of R&D can be utilized effectively. Thus engineering, production, and marketing departments have to be educated to accept and use R&D efforts. This provides the pull necessary to introduce R&D into the system.

Given these conditions, Unilever officials consider it logical to follow the same road in LDCs that it has elsewhere, fitting R&D activities to commercial needs and expanding with the market.

NOTES

1. The multiple controllers include the head of the Research Division, the heads of the labs themselves, the Product Area Managers, and the Research Project Managers, each having some overlapping jurisdiction as a result of their responsibilities to different units within the company.
2. The pressure arose from a new company official who had been a science advisor to the Canadian government and became head of the development lab. He pushed for ties with Central Research, which was readily accepted by the Chairman of the Canadian company, who also sought a wider R&D program.
3. To support these personnel, the lab contains a wide variety of biological, chemical, and electronic facilities. The lab also maintains a research library, which in 1976 had a purchasing budget of £44,000, distributed £28,000 to scientific journals, £6,000 to books, and £10,000 for journal borrowing facilities with other libraries. These materials cover a wide range of subjects including journal abstracts, analysis, biochemistry, chemical engineering, computers and statistics, corrosion, ecology, and pollution, engineering, general chemistry, information science, instruments, life sciences, management and commerce, organic chemistry, physical chemistry and physics, polymer chemistry, research management, safety, skin, soaps and detergents, surface sciences, textiles, timber, water, and many others. These resources come from many nations in several languages.

   About one-third of the potential users of these materials are in the library each day to consult various sources. Many of these resources would be necessary to support even a specialized lab, because there are certain "core" scientific journals that all scientists need to consult no matter what their discipline or techniques. In a laboratory working only on hair, for instance, biologists, chemists, physicists, analysts, and data processers would still need to be supported by adequate literature.
4. Being a major lab within central research, PS generates a large number of research reports. It circulates every three to four weeks a current abstract, *Research Abstract*, covering some 25 to 30 percent of abstracts received from all labs plus all of those that it produces. Within Port Sunlight, some

thirty to forty professionals receive these abstracts and read selective parts of it. In addition, the Key Word and Content (KWAC) system is available to all the labs on a computer printout and microfiche.

The PS library dispatches many of its basic reports automatically to other labs, but most are sent on request. There is an additional flow of reports from professional to professional, but the lab library and the information department have no knowledge of the scope of such exchanges, nor of their actual use. A cursory review of citations made of prior reports of other teams showed that about 90 percent of the citations were to reports from the PS lab, with the remainder from their central or affiliated labs.

5. Coordination between the lab at Isleworth and the U.S. labs appears to be strong and consistent if one looks at the reports and trips, but an examination of the actual flow of technical information would show that exchange as erratic. The United States is the largest single market for toilet preparations, and toothpaste comprises by far the largest portion of TP sales; the U.S. company has a higher profit ratio than in the U.K. and therefore more funds for product development, though it has fewer products in the pipeline. Cooperation has been improving lately because the United States has felt it had something to offer and saw itself in a more equal negotiating position.

6. An official at Colworth stated that the most heavily published sections are basic research and toxicology, which have a wide clientele. Applied research sections publish less; commercially oriented sections are stopped by security problems and gain less personal satisfaction from publication. However, there is a lab policy to promote publication.

7. Melanin, the major pigment of skin, is synthesized by a specialized cell called a melanocyte and then dispersed in the cell layers. Both Caucasians and Negroes have the same number of melanocytes, but their color differences arise mainly due to the way in which the melanin is dispersed.

8. Professor J. Boldingh, October 13, 1975.

9. All officials indicated that if the communication had to be over 1500 miles, it would be hard to handle, especially because the "man in the kitchen" would be very busy trying to carry out his activities and decipher and apply the information given him. The further away the lab is geographically from the center, the harder it is to keep communication effective, simply because visits would be cut to the bare minimum, if not to zero, and it would be difficult to narrow the gap in perception.

Johnson & Johnson: R&D in Health Care

INTRODUCTION

The primary products of Johnson & Johnson are in the health care field, including such items as toiletries and baby care products. The company's activities fall into three major categories: Health Care (domestic), including drugs, diagnostic products, surgical dressings and instruments, sutures, toiletries and hygienic products, and veterinary products; Industrial and Other (domestic), including products for textile, agricultural, aircraft, automotive, building, food, health care, paper, and other industrial markets; International, encompassing any of the products from the two domestic operations that are relevant for the overseas markets and not marketed in the United States.

Total sales were approximately $2.5 billion in 1976—50 percent in health care, 10 percent in industrial and other, and 40 percent in international sales. Over and above quality control and engineering development costs associated with manufacturing, research expenses topped $100 million in 1976, or about 4 percent of sales. Over the past several years research expenses have been larger than those incurred for advertising worldwide.

The name Johnson & Johnson is used on only three of the

Johnson & Johnson domestic companies—Baby Products, Dental Products, and Domestic Operating Company. Thirteen other major company names and trademarks in the United States relate to specialized products for hospital use, nonwoven textiles, sausage casings, industrial tapes, feminine hygiene, and a variety of drugs. Each of these companies has its own R&D activities, but not all of them are extended overseas.

The company began operations in New Jersey in 1886 and established its first laboratory, in bacteriology, in 1891, in an effort to obtain a sterile hospital dressing. In 1897, it improved the sterilizing technique for catgut sutures, and in 1899 developed and introduced a zinc oxide adhesive plaster.

It acquired a Canadian company in 1919, and the first overseas affiliate was established in Great Britain in 1924. Since then affiliates have been established in more than forty countries, expanding not only through the establishment of new facilities but also through acquisitions both in the United States and abroad.

The management philosophy of the founder, General Johnson, was that of decentralization and independence among the various company units. There is, therefore, no central research and laboratory for the company as a whole. Each of the companies has set up whatever facilities it deemed appropriate for itself. And each of the foreign affiliates is left to determine the nature and extent of R&D activities suitable for its objectives. This is not to say that some gentle prodding has not occurred at times, nor suggestions emanated from the parent headquarters. But in many cases no mechanism exists for coordination of research activities worldwide, nor is a "corporate lab" responsible for R&D across the many companies. Each company lab is a self-contained unit, though all are expected and are found quite willing to pass along any information desired by an affiliate for whatever use it wishes. This exchange occurs without a separate "fee for service" (save for direct lab costs, when required to answer questions), despite the possibility that production in a foreign market may cut off exports from the company sending the technical information out of its R&D lab. Johnson & Johnson does require, however, that, when legally permitted, all international affiliates pay a general fee based upon a percentage of sales (averaging 2-3 percent of sales) to Johnson & Johnson International. For this fee the affiliate is provided, either directly from Johnson & Johnson or indirectly through U.S. or international affiliates, whatever product technology and assistance it requires in the conduct of its business. However, because of management considerations, Johnson & Johnson International does not pass on any portion of this fee to the U.S. labs.

In order to make this study manageable and still illustrate the problems of international research operations, it was decided to focus on health care and baby products, since these tend to be the first ones produced by an overseas affiliate. However, where interesting illustrations were found in other product areas overseas, these have been drawn upon. Visits were made to the research center for the domestic operating company, which is responsible for the major line of health care products, the baby products lab in the United States, and the counterpart labs in Canada, Britain, and Brazil.

U.S. LABS

Until about five years ago, the Domestic Operating Company (DOC) contained not only the Health Care, Patient Care, and Dermatology Divisions, but also Baby Products, Dental Products, and Surgikos. Then these last three split off on their own and carried with them their product development activities. The former DOC lab was split up, and the divisional labs remaining within it wanted to report more directly to their operating divisions. This was accommodated through a new organizational structure, but it was also considered necessary to maintain an exploratory unit that would do some pioneering research for all of the divisions.

The missions of these separate labs are sufficiently distinct, and their management sufficiently separate, for their roles in the DOC to be seen as rather different by the managers of each. Obviously, their perceptions are affected by their own view of their activities, as distinct from those of others, and as distinct also from the view held by top management of the total R&D activities of the DOC. (It is a well-known feature of this type of research, relying on perceptions of individuals regarding their roles and the activities of their business units, that views are sometimes contradictory, or at least not parallel. Where decentralization occurs, these divergent views may be accurate, since the units are in fact themselves diverging. In other cases, divergence in views results from confusion over roles. No judgment of the underlying causes can be made without extensive interviewing of a number of management personnel and levels, which was not feasible in this particular study.)

More recently, reflecting the continuously changing demands on R&D, the structure has been shifted again—this time towards more centralization. A Vice President for R&D of the DOC is now a member of the Management Board of the DOC and has the R&D Directors of each Product Division reporting to him. These divisional labs are responsible for developmental research in their product areas.

Basic research, which may merely lead to "something, sometime in some area", is not done in the DOC labs. Exploratory research, information leading to specific products or solving specific problems that may arise somewhere within the product line, is the responsibility of the Exploratory Research lab. This lab feeds ideas on new products to the divisional labs for further development. It may also pursue product ideas outside the interest of any division, hoping to catalyze new departures by one of them. If none picks up an idea it considers worthy, it may take the item through the development stage to an actual product for field testing. This was done in a digestive aid for cystic fibrosis, which Baby Products later picked up. Since certain of the divisional labs are too small to maintain their own exploratory research units, the exploratory lab of the DOC also supports these companies and receives some of its budget from them.

The market-orientation of product development is reflected throughout the entire corporation, with the R&D program control being strictly the responsibility of each division or company. Given the variety of products in each, the R&D program tends to emphasize the development activities rather than research. These labs remain fairly separate in their orientations and activities, with little program coordination among them. A council of R&D directors of the U.S. labs meets five to six times a year, but it focuses principally on problems of research management.

An international R&D meeting is held every three years, attended by professionals and managers from each R&D lab. Each domestic lab makes a presentation of selected aspects of its program. The major benefit of such meetings is the development of face-to-face contacts and personal relationships that ease the daily flow of information. After the meetings, personnel from overseas labs will frequently visit their counterparts in the United States and Canada. The meetings have become so significant that they are attended by many officials other than those directly involved in R&D.

At present, Johnson & Johnson has an International Quality Assurance group that undertakes a monitoring and auditing function on testing of overseas products. It ascertains that quality control procedures are adequate and appropriate, and sees that initial runs are properly evaluated. It reports on product quality vis-à-vis local governmental standards and also correlates product quality with U.S. company standards. Liaison officials are detailed to provide assistance whenever an overseas affiliate asks for help. A major problem is the unavailability of raw materials in certain overseas companies, and the International Quality Assurance people offer much assistance in their evaluation.

DOC Labs

The Domestic Operating Company has three major product divisions, with a development lab tied to each. These, plus the exploratory lab, are housed in a single complex under a single administrative unit, which is also responsible for technical service and quality control in the entire company. The organizational structure is shown in Figure C1.1.

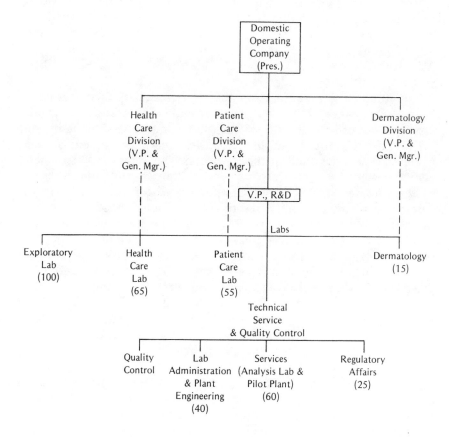

Figure C1.1. Organization of the Johnson & Johnson Domestic Operating Company

Exploratory Research. The exploratory lab was formed seven or eight years ago, but some of the work it is now doing was accomplished by bootlegging time of scientists from the various disciplinary labs beginning about twelve years ago. The necessity to do this bootlegging led to the decision to form this research unit. Coordination with these other divisions is achieved through two-day meetings at least twice a year to explore divisional needs and the work of this lab. An advantage to this setup is that the development labs are an interface between the exploratory lab and the marketing divisions of the company. They provide the necessary pull out of research into development and commercialization. The scope of the lab's activities is illustrated by the types of personnel it has, its budget, and the specific projects it undertakes.

The lab employs somewhat fewer than 100 people, encompassing professionals from biology, microbiology, veterinary medicine, bioengineering, physiology, biochemistry, organic chemistry, histology, pharmacology, pharmacy, polymer chemistry, and rheology. About half of the unit's personnel are Ph.D.'s, with virtually all the rest having at least an undergraduate degree.

The funding for exploratory research comes from allocation to it by four companies—DOC, Baby Products, Surgikos, and Dental Products, all previously part of the same company before being split off. Their contributions normally run less than one percent of sales, which is a charge against their profits. The research unit usually justifies each of the projects it is undertaking by obtaining approval of the presidents of the four companies. Within the DOC, for example, the charges are allocated to each division, whose management often sees a conflict between R&D goals and their own objectives, which are more short-range and do not readily include introduction of completely new products and longer-range product planning. For example, the exploratory lab spent $1 million to develop a digestive aid that will go into the Baby Products Company to help with problems such as cystic fibrosis, and may be picked up by both Britain and Brazil and adapted for their own marketing needs. However, none of the expenses borne by Exploratory Research were offset by returns from these affiliates.

Projects. Less than half the exploratory unit's program is in gathering scientific information to provide innovative leads for product development. The other half is for product development activities themselves aimed at discovering new business opportunities inside or outside the divisions' present lines, but still within the health care area. This kind of effort led to the development of the

Dermatology Division and the expansion of dental products into a full-fledged company, both placed outside the then-existing product divisions.

To make certain that the products they are examining fit within the company, the unit has a small market research group that examines the possibility of sales and the way in which the product would fit into existing lines. And a computer program within the research unit tests out the return on investment from any given opportunity uncovered.

The projects undertaken by the exploratory group cut across several product lines. For example, the project on skin is looking at wounds, burns, inflammation, dryness, and the penetration of skin by lotions or oils, plus the effect on skin of various adhesives. A joint project was set up between the exploratory lab and the labs in Health Care and Patient Care on adhesive tapes, with the exploratory unit working on polymers, synthetic adhesives, and the biology underlying adhesion, the Health Care lab working on different adhesives, and Patient Care lab working on varieties of tapes. Monthly meetings of the professionals assigned to the project in each lab provide adequate liaison. Each lab pays its own cost of the work done, and each has picked up that portion of their work that is most relevant to its market opportunities.

Coordination with Overseas Labs. Overseas labs are equally independent of any control from the United States, so that communication from any of the U.S. labs is through a formal exchange of information or personal visits to see what might be of interest. semiannual reports are written on each research project—twenty-five such projects within the exploratory unit—and these go to some fifty-two other labs, including many of the overseas labs. These reports are received by the Director of Research, who has the technical ability to determine whether there is anything of interest in them. If so, the directors will write back to the exploratory unit for additional information, since the report is primarily descriptive, with little or no actual technical data. More detailed information will be sent to the lab, including all that is necessary to put the affiliate in production, if that is what is wanted.

These formal reports are supplemented by two-way visits from the United States and to the United States from abroad, with the frequency related to geographic proximity. The most frequent visits are between the United States and Canada, with about two such visits exchanged with Britain and occasionally with other

European labs. Lab officials indicated that more frequent visits would be desirable, but they are expensive.

The value of this outward communication from the DOC labs is evidenced in that most overseas products in these fields have stemmed from the DOC output. A few modifications have flowed back from Australia, Britain, and Canada, but the information flow is still generally outward from the United States. Once the personal contacts have been developed through visits, the professionals tend to deal directly with each other, frequently by telephone. No formal procedure has been set up to channel communication with a foreign lab, outside of regular written reports.

Health Care Lab. The Health Care Division produces consumer products—first aid, oral hygiene, skin care, internal sanitary items, and the like. The R&D Directors of the DOC report directly to a Vice President for R&D. Reporting to the R&D Director, at the time of the interview, were four section managers having responsibility for clinical research, first aid products, oral hygiene and toiletries, and proprietary and athletic products. Some sixty-five persons work in this lab, excluding secretaries and administrative support. An average ratio of one scientist to one technician exists, not including the managers and project leaders, who are also scientifically trained. Each project leader is responsible for several projects simultaneously, with a senior scientist in charge of each separate project.

Projects. A new project in the Health Care lab can be started anywhere, but it occurs mostly from within the lab itself at various levels. Once an idea is initiated, there is a meeting within the lab and a project form is written up including the goal of the project, the type of project—whether applied research, product development, product adaptation or improvement, or whatever; the technical feasibility of the project, the marketing rationale, the timetable for completion, and the proposed budget and equipment needs. This proposal is then approved by the Director of R&D, the General Manager of the Division, and the brand manager to whom this particular project would fall. Once approved, the project is assigned to a leader, based on the major technology involved; he will obtain cooperation from other units as needed.

A "critical-path book" is kept on each project, reporting on a quarterly basis the accomplishments to date and the objectives sought in the next three months. This report goes to the marketing group, and a meeting is held with Marketing on all such quarterly reports. These sessions act as a "early warning" system for Marketing to inform them about what will be coming out, so they can prepare to commercialize the products. As the project is finished, a "fact book" is written on all aspects of the product development—research, development aspects, clinical safety and efficacy studies, specifications, pilot tests, and results of the plant run. (The lab is responsible for the first few runs of production on plant scale, to ensure that the technology is fully checked out.) At the pilot product stage, a consumer test is run with a "blind product" comparison against a competing product to prove that the new item can stand on its own quality without advertising. The fact book includes the results of these tests and also provides backup for prospective advertising claims, which Marketing must then follow. This final writeup is a very complex piece of work, that has to be okayed by clinical research, technical service, and those concerned with quality assurance and regulatory affairs. These triple checks reduce risks considerably. The fact book is the basis for the transfer of information to overseas affiliates, since it has all production information in it. Any foreign laboratory intending to introduce the product can get this book merely for asking, putting its company directly in the business, though there are obviously security restrictions on the distribution.

The number of projects leading to new products is dictated by the ability of Marketing to pick up new product developments and commercialize them. If they are constrained in their ability to introduce new products, then the R&D program would shift to cost reduction rather than expanded sales. However, the lab does not concern itself with process development, which is the responsibility of technical services.

Outside R&D Support. Although the activities of the lab appear to be largely self-contained, some ties do exist with neighboring universities, such as Rutgers and Princeton. The research at these universities is different from product development, but it does provide some background information useful to the lab, and helps the company keep up with the scientific developments of the day.

In addition, the lab is kept abreast of industry positions on R&D issues through associations with the National Institute of Health, the National Academy of Science, the Pharmaceutical Manufacturers Association, the Cosmetic, Toiletries, and Fragrances Association, and the Proprietary Products Association. These groups are useful in an exchange of ideas on the state of the art, but again not on specific research projects. Attendance at local professional meetings (in New Jersey, New York, and as far away as Washington, D.C.) is encouraged, and each professional scientist usually gets to go to one national meeting per year within his professional discipline.

Communication with Overseas Labs. The overseas labs are not provided *continuing* support from the Health Care lab, and few of them have asked for troubleshooting assistance. No formal exchanges of professionals among the overseas and U.S. labs have occurred in the Health Care Division, but frequent visits occur between the Canadian, U.K., and U.S. labs. These visits provide personal contacts to ease communication on specific projects.

These personal contacts led to a cooperative project between the United States and the United Kingdom when a U.K. lab manager picked up a project in the United States that had been rejected, then successfully completed the phases that had stumped the U.S. lab. On a visit to the United Kingdom, the U.S. lab director noted the results obtained there, and brought them back for testing in the United States. The two labs instituted a process of continuous exchange of results, leading to successful product development. Lab managers feel that this kind of cooperation could be usefully expanded.

Patient Care Lab. Since the division markets products in the areas of orthopedics, tapes, thermometry, absorption of body fluids, and wound dressings, the development laboratory emphasizes the development of new and improved products in these areas. Additionally, however, new product areas are continuously evaluated for viable business opportunities for the division.

The lab's organization includes not only project leaders in each of the above areas, but also clinical research and technical support for sales and marketing. Of some forty-seven employees, thirty-one are professionals from the fields of polymer chemistry, textile chemistry, microbiology, synthetic organic chemistry, and analytical chemistry. In addition, a biomedical information and management information center provides mathematical modeling for projects on an as-needed basis.

Projects. The work of the lab is aimed primarily at development and technical support. Developmental research undertaken by the lab is aimed at a clear need, such as the development of a replacement for plaster of Paris. Specific activities include the demonstration of technical feasibility, stability of the product, processes of manufacturing, and proof that the product is clinically efficacious.

Technology expertise has evolved within the development labs to the point that overseas labs have from time to time called upon specific individuals for technical assistance. Examples of such areas are surgical tapes and plastics processing.

Community Relations. The success of the Patient Care lab is significantly dependent on its relationships with the outside science community. Contract research has been provided by Batelle Institute, by Rutgers University labs in orthopedics, and by others. The lab spends between $100,000 and $200,000 each year in clinical research on new technologies, and it retains over thirty consultants each year on various assignments, extending from hospital needs to space science. Lab officials assert that they could not hire this expertise if they had to obtain it on a permanent basis within the lab, and that without such assistance several of their projects would simply be stymied.

Some of the results from the lab are published, as illustrated by the publication of a book of collected research in the past four years, plus four or five articles in the same period. Given the developmental nature of the research, the results are not readily disclosable, and probably would not fit within the scope of most professional journals. In any event, there has been little effort to push for publication of lab results. A better indicator of the capabilities of the lab is the patents that have been obtained as a result of its work.

Communication with Overseas Labs. Communication with overseas affiliates is mainly from the U.S. lab to those abroad, and is similar in form to that in the areas discussed above. The U.K. lab sends a professional at least once a year to New Brunswick, but the geographical distance makes it difficult to visit frequently. The interchange is most useful for R&D management, in particular, when research projects are ongoing in similar product lines. Some customer needs in England remain different from those in the United States (for example, disposable thermometers are not needed in British hospitals, because temperatures are taken only twice a day), but there are numerous areas in which the products are virtually the same, making information exchange quite valuable. However, the pressure of daily workloads sometimes makes lab professionals reluctant to take the time to answer questions from abroad in detail. Some directives and incentives need to be given them in order to induce more ready response to overseas needs. One such method has been to hold a worldwide meeting of hospital product research directors in which each reported on his program, stimulating the exchange of ideas and problems. These personal contacts have produced greater attention to requests among labs. However, given the differences in interests and orientations, there have been no joint projects with overseas labs, although there have been a few exchanges of personnel. The level of R&D activity in patient care overseas remains considerably below that in the United States.

It should be noted that differences in approaches and product lines have resulted in the absence of significant joint projects across the development labs even within the DOC itself. Recently, however, the tendency toward further decentralization within the R&D divisions of DOC has been reversed, reintroducing greater coordination into what had become a fairly competitive system overall. In summary, it appears that the overseas labs will remain *relatively* isolated from their U.S. counterpart mainly because of geographical distance and infrequent personal contacts.

Technical Services. The technical services units provide a variety of services to the other labs as well as to the product divisions. They also provide the administrative support for all the labs and maintain the library that includes computer printouts of abstracts, patent searches, and other sources of scientific information. They also run the pilot plants for polymer processing and chemical engineering, and include an engineering development group for design of new equipment.

Centralization of services for the labs is important as a cost-reducing factor, but as labs grow, they can take on their own service units. However, labs' sizes are limited by the increased risk that ideas, people's motivations, and equipment may get lost in the system; the breakdown in communication becomes significant as labs increase in size and units become distant from each other.

Technical service groups are located in each of the manufacturing plants and report to the manufacturing division. These groups:

Provide process and material support for existing products (i.e., troubleshooting).

Develop and implement cost-improvement programs based on technological change and innovation.

Evaluate and have approval authority for any divergence from specifications of materials and/or processes.

There is a separate National Technical Service group also reporting to manufacturing that audits and critiques the programs of the local technical service groups. In addition, this group is responsible for:

Controlling and directing the transition of programs from Research to Operations.

Identifying and initiating advanced technology programs.

Providing consultative technical input to the individual manufacturing plants and to subsidiary and international companies for products and processes for which the DOC is the originating company.

These groups, acting in concert, are the established channels for communication between Research and Operations.

Communication with Overseas Affiliates. The structure of Johnson & Johnson R&D labs leaves each operating company virtually autonomous in the determination of the product line and the way in which it gestates. Given the marketing orientation of each of these companies, lags often arise between the development of new products by R&D in the United States and their adoption overseas. The local marketing divisions tend to be slow to pick up the products, taking a conservative view of their funds and time.

Since, in Johnson & Johnson, the main source for new products for any of the foreign affiliates is largely from among those marketed in the United States, the major problem is that of communicating to the affiliates what is available and how useful it might be in their own market. The decentralization of the company permits lags in communication to exist simply because there is no strong pull from the marketing units in the affiliates overseas.

Some over-the-counter drug products are so well standardized that they can be produced overseas without adaptations. For these, there is an apparent need for improving the present channels of communication between the U.S. scientists and overseas labs. The major problem seems to be that of identifying market needs, which is not yet adequately done in all the overseas affiliates, and for which the U.S. labs have no facilities either.

In sum, Johnson & Johnson is still discovering many of the problems of handling international R&D and product commercialization and ferreting out the answers—as will be seen from the review of the development of the British, Canadian, and Brazilian labs in subsequent sections.

GREAT BRITAIN

The British affiliate was formed in 1924 to produce baby products. It had a few people responsible for quality assurance, but no research activities of its own. In 1954, the baby products operation was moved to Portsmouth, with the addition of an elementary lab, which was involved also with adhesives.

In the late 1950s a central lab was created at Slough, at corporate headquarters, to do some basic research in organic chemistry. A director for the lab was hired from Scotland Yard. He was mostly interested in academic-type basic research, such as in absorptive capacity and talc lubricity. He hired half a dozen Ph.D.'s, but the lab remained technologically remote from manufacturing and did little on existing products. It operated without relation to the technical needs of the plants, and its output was not well appreciated by the board. This central lab had little contact with the plants except through various assistants to whom the technical service units at the plants reported on their adaptations of products and processes received from the United States; there was very little feedback from lab scientists on these reports. The lab was closed in the early 1960s, and its director departed. The principal scientists left for positions in other companies, and their assistants were placed in facilities connected with the plants.

This original lab orientation arose out of the director's feeling that the British could "do anything the United States could do and better", and since the U.S. labs had been begun only five or six years earlier, he thought he could catch up to American expertise. But the lab remained aloof from the product and the market, and was too costly compared to the size of the company at the time. Since it generally takes at least five years to get any totally new product through the R&D process, J&J officials assert that a lab must be closely integrated with the rest of the company, or else top management will consider it to be a "black art" with unmeasurable and possibly undesirable results.

Present research activities, therefore, grew out of the quality assurance (QA) and technical service (TS) operations at the plants. The two activities had separated in responsibilities; those doing technical service also made the product and process adaptations necessary as the items were picked up from the United States.

Gradually, under pressure caused by the company's growth, some product development activities were split off from the TS group, using those who were capable and interested. Scientists have to develop an expertise and knowledge of how to develop products, which can be aided by TS experience in the plants.

The TS group was pressed from the beginning by the Marketing Division to satisfy the different tastes in Britain and therefore to make product adaptations. Since the Manufacturing Division wanted more technical service, the functions had to be separated. Initially, the Marketing Division was given responsibility for QA & TS, but in order to clean up direct responsibilities and to slant the TS group towards manufacturing problems, the budgets were separated in the mid-1960s, with product development activities remaining the budgetary responsibility of the Marketing Division. As the product line has expanded, all QA-TS and product development activities have increased. Technical service has been increased to customers who have had problems with the products—such as may occur when an affiliate of another U.S. company in Britain changes the ingredients of its British product, only to find that the J&J component it is buying performs differently. The two U.S. companies can be helpful to some extent, but frequently the two British companies, both using different ingredients, will have to work it out on their own. (In fact, this independence is sometimes an asset: company officers feel that they are able to recruit good researchers on the basis that they will be doing something for the British company and British consumers—rather than working for the parent company in the United States for worldwide customers.)

As the number of products has increased, the number of different customers has also increased, and more time has been demanded for customer technical service. Both QA and TS need to be located at the plant but not reporting to the plant manager. (United Kingdom law forbids the QA unit to report to the plant manager if a license is required to produce those particular items.) However, product development activities should, according to J&J managers, be more closely related to Marketing and top management than to Manufacturing. This eventually permits a separation of R&D activities from the plant as they become more sophisticated and development oriented.

Organization

J&J has four manufacturing units in Britain—baby products and adhesives at Portsmouth, surgical products at Gargrave, nonwoven fabrics at Pontllanfraith (in Wales), and sterilized products in Slough. There is a corporate Director of Research for the entire British company, having responsibility for QA and TS as well as microbiology, product development, and medical developments. A QA and TS unit is attached to each of the four plants; each also reports to the Group QA and TS manager, who in turn reports to the Director of Research. Microbiology managers are located at Portsmouth and Gargrave, with the Portsmouth manager being responsible for the other two plants as well. Three product development managers (at Portsmouth, Pontllanfraith, and Gargrave) report to the Group R&D manager, who is located at the Portsmouth lab near the plant. Within the Portsmouth lab, separate divisions work on baby products and toiletries, new product development, and health care (principally adhesives) at Gargrave. Two other units work on fibrous products and those coming out of Surgikos in the United States.

The final unit is at Slough, under the Medical Advisor, who is an MD, with a staff covering toxicology, clinical trials, and a medical library.

Capabilities

Some twenty-five professionals cover the five major research areas, supported by about twenty technicians and a number of other administrative and support people. Just under half the professionals have Ph.D. degrees in the areas of microbiology, biology, and zoology, chemistry and pharmacology. Other disciplines in which technicians have been trained include chemical engineering, physics, and textile science.

Upon entering the company, the new researcher is given a short acclimatization course and then put to work on the bench. There are no in-house courses by which to educate him further. However, scientists are encouraged to take courses in local schools or universities, attend symposia, and take some management courses. No structured training program for each of the professionals has been developed, though the company did sponsor the postgraduate study of the director of research, who joined as a research chemist once he had completed his Ph.D.

Some training occurs through personnel transfers, though not on an exchange basis with the United States. A microbiologist was sent to New Brunswick (N.J.) for a year, a research manager for one to two months to New Brunswick, and a research chemist to Canada for two months. Many others have been sent over for shorter periods. All were enriched by the stimulation of interests in the U.S. and Canadian labs, from which a great deal is gained simply because there are similarities in research techniques and language. In addition, exchange of information is facilitated by joint projects that last for several months, with each lab working independently on different pieces of a given project. Lab officials feel, however, that more could be done both through visits and joint projects to enhance the training of professionals in Britain.

Few publications come out of the British lab, but some evidence of their capabilities is reflected in the three or four patents that are obtained each year.

The product development budget amounts to about 2 percent of sales; the QA-TS budget, which is part of manufacturing costs, also amounts to about 2 percent of sales. Annual capital expenditures and building maintenance amount to £136,000 for the three labs together. Expenditures for the library come to £2,000 annually, supplemented by free assistance from university and public libraries and the National Lending Library Service, which will reproduce any article published anywhere and send by return post.

There is no rule of thumb on the amount of expenditure for product development as compared to sales of the company. In fact, baby powder sales supported all R&D efforts in all fields for some time, and yet no research was done on talc or powders in the United Kingdom. There is, therefore, no fixed ratio of development effort to sales; the budget is determined by a proposed set of projects charged to the corporation as a whole, and met out of corporate funds, which are then allocated back to the several marketing divisions. A small amount is assigned to the corporate research unit and paid out of total funds.

Laboratory facilities include an engineering shop to produce machinery for the product plant; each lab has its own machinist and chemical engineers and designers. The scope of the lab employees' interests and capabilities is indicated by the areas in which they work. Portsmouth does work in plastics, adhesives, polymerization, emulsions and foams, toiletries, surfactants, biochemistry, and microbiology. Pontllanfraith works on fibers, nonwoven fabrics, and polymeric lattices. The medical affairs unit at Slough is concerned with medical research and toxicology, but also advises on regulatory affairs and on toxicological research at each of the plants. Several of the projects are split for control and evaluation among the various labs due to the crossover of their interests and expertise.

In addition, the labs have some facilities for doing consumer research, especially in hospitals, but by and large this is left to the marketing division. It sometimes hires outside consultants to do consumer research, and the R&D lab may accept the results or not, depending on the research design and the way in which it was conducted.

Projects

In the early postwar period, the British plant adapted and modified the U.S. product lines. However, a few affiliates did develop some new products that were not in the J&J line—for example, disposable syringes and needles. These were sold under a name other than J&J until the United States decided to produce this line. (Company officials stressed that, contrary to some public images of company decisionmaking, events do not happen in the standardized, routine, textbook fashion; happenstance and unique circumstances are more determinative than rational planning processes.)

In some instances, the British company has gained some ideas, concepts, and methodologies from examining the "failures" in the United States. In one case, a chemically-defined diet called Jejunal was developed in the U.S. for patients prior to and after surgery. Since the intake is completely used by the body, there is no residue, which permits people to eat before surgery rather than entering in a half-starved state. But the general lack of education in the medical field concerning patients' nutritional needs, especially under the stress of an operation, prevents doctors from understanding the desirability of such a diet. The product was attempted in the U.S. market, but the marketing division was pessimistic, since it was seen as too specialized, thus requiring a different marketing technique and approach. After considerable sales effort, it was dropped as economically unfeasible. The U.K. company saw the diet as a means of helping patients with kidney problems by reducing the load on the kidneys. Research on the product opened up a new field in chemical nutrition, and the possibility of success in diseases such as cystic fibrosis. A whole new science field is just beginning in nutrition, and the British company sees this as a means of getting into that area.

In a second case, the British company developed a foam pad for "Band-Aid" adhesive bandages, that the United States had unsuccessfully tried to develop. The British company picked up the idea because it wanted something to change the marketing image of the bandage and saw this as a fit within its own "market brief" for the sales of "Band-Aid." It succeeded, using appropriate materials that were not available to the U.S. company, but were in the United Kingdom. The success of this item illustrates the necessity for Marketing to have a clear perception of a need, or else R&D finds itself pursuing a product that Marketing is unwilling to pick up.

J&J managers have drawn the conclusion that local R&D activities above the QA-TS level should start from market-oriented needs and with the affiliate's independence in serving its own markets (both domestic and export). Otherwise, a local lab should not be established, and the affiliate should rely on the R&D of the parent company. Only if the markets in all countries are basically similar would it make sense to allocate to affiliates pieces of research projects directed and controlled from the center. Too often, chance elements help to dictate the research projects required, and these cannot be discerned from the center. For example, competition in the U.K. adhesives market forced a greater attention to R&D in this area, in addition to the research director's background being in adhesives (a coincidence that helped dictate the program of the lab).

One criterion dictating the nature of the projects is that the U.K. company seeks to reduce its costs by using locally available resources to reduce foreign exchange needs.

But the primary factor in determining the nature of the projects is the market differences between the United Kingdom and the United States. In the consumer lines, many tastes in Britain are different, and the segmented markets in the U.S. are not likely to be replicated in national markets abroad. The company must test in each case what market will take which products and what adaptations are needed. For example, the size of the package must be different, because the price a British housewife is willing to pay for a single purchase of baby products, toiletries, adhesive bandages, etc., is less than in the United States. Frequency of use is different, since elastic cloth bandages are used more frequently in Britain; there is a lower per capita use, meaning distribution must be different; and less advertising is done. A paper carton package was acceptable for adhesive bandages in Britain, while the U.S. company keeps the metal box. In addition, diaper liners sold better in Britain than in the United States as an intermediary between disposable diapers and washable cloth diapers. To meet the competition in bandages, the U.K. company had to develop a cloth-based elastic bandage, which the U.S. company had not considered useful.

In the professional products, however, especially for hospitals, the United Kingdom has taken almost all of the products directly from the U.S. line. Doctors, for example, know what their peers are doing, and therefore the markets tend to be similar. In this line it would not make much sense for Britain to develop substantial laboratory capabilities in new product development.

In the nonwoven fabrics business, the line has been expanded beyond or in advance of the United States into sponges, as a result of extensive consumer research. In addition, the United Kingdom picked up a development by the Dutch lab of J&J of disposal tableware from nonwoven fabrics.

The development of projects undertaken by the labs begins with a research proposal, which includes a marketing plan and comparison with U.S. experience in the product area. The proposal goes to the marketing manager, and if he approves, it goes to the R&D manager at Portsmouth for his assessment. If he has checked it out successfully, it will then go to the Director of Research for final approval. If need be, he will give it further assessment, possibly checking again with the United States or Canada. Such correspondence would go through product groups or to the exploratory lab at New Brunswick. After the product has been developed and appropriate adaptations made, the responsibility of the lab extends into the plant, where the first several runs are the responsibility of R&D, including the test marketing and launch. R&D labs recoup some of the costs of these first runs from sales to the Marketing Division if the test markets are successful. The lab remains responsible for cutting waste in production down to standard levels.

The U.K. company also has responsibility for serving a number of the present and former Commonwealth countries, with which it has had long ties. The U.S. parent companies did not develop the products for these countries; they were served out of the United Kingdom, which adapted U.S. products for use in those markets. The United Kingdom does product development for Africa, the Near East, and Asia, not only making changes in products for export but also in some produced by affiliates abroad. Thus, the Indian affiliate of J&J gradually developed its own capabilities and, through the necessity of a ban on imports of materials, developed a lab to incorporate indigenous ingredients.

Also among the projects undertaken by the labs are regulatory activities. Some of these activities are made easier by U.S. FDA standards being understood by other countries, so that when the labs show that these standards have been met, the products are frequently approved locally. Still, substantial time is required by the lab to meet local regulations. Some of the increased time is required by the European Community's becoming more concerned with developing regional standards in addition to those imposed by national governments. About 10 percent of the labs' time appears to be spent in regulatory affairs, though any new business development requires about 15 or 20 percent of the time to be spent meeting with governmental regulatory agencies.

Coordination with the United States

The first responsibility in the development of any new research proposal is the examination of the J&J activities worldwide (as well as related information from outside the company). The lab scientists screen the reports from J&J to see where the company might position any U.S. products in the British line. Given the wide range of possibilities for new consumer products, the proposed projects generally are more ambitious than the lab can accomplish in any short period of time. A tradeoff has to be made between reducing the number of projects or possibly reducing the chance of success in a larger number of projects. Higher priorities are given to those which seem to provide a greater chance of success, with more resources being committed to them. For example, the British lab now has a man in Canada learning about nonwoven absorbent technology for use in "J-cloth," tea bags, and pulp stabilizers. Sometimes, the lab gets ahead of Marketing or Manufacturing, as at present when it has taken a successful nonwoven product through the pilot project stage, but Marketing and Manufacturing are not yet ready to make the necessary capital expenditures. For maximum effectiveness, a nice balance is required between R&D activities and the readiness or capabilities of other departments in the company.

Since the U.S. labs are totally geared to the U.S. market and the U.K. labs are looking after their own market, a ready fit does not always exist in the division of responsibilities in a single project. The U.S. labs are willing to pass on any results to their U.K. colleagues, but they are seldom ready to work directly on a U.K. problem. Even so, it is extremely difficult for the R&D managers in the various companies in J&J Ltd. to conceive of how they would operate without reliance on the labs in the United States, making it virtually impossible to calculate the value of the contribution they receive from the center.

In the case of the development of the "Clear" adhesive bandage, reports of the various labs showed that the U.S. and U.K. labs were working along parallel routes, so a joint project was set up, with each lab having its own project director responsible for coordination and each specializing in pursuit of different routes to find answers to the same problems. Visits were made between the labs to maintain coordination. The United Kingdom was more successful, as it happened, since the necessary materials were available to them and not to the United States.

This case again illustrates the necessity of personal visits to provide necessary communication and to set up appropriate cooperation. Visits of U.K. lab personnel are primarily to the U.S. counterparts, but they will also go to Canada since the operations there are on a scale similar to that in Britain. One professional is in the United States at least every month or so, as needed, with approval of the director. Section leaders visit the U.S. labs at least once a year, and visits to European facilities are made frequently without the director having to approve these. Visits are also made to Africa and other countries for purposes of teaching others about the adaptations in new products. The frequency of visits to the United States has increased substantially since the early postwar period or even the establishment of the lab some fifteen years ago. Even at that time, trips were made more frequently by ship than by air, increasing the time and cost relatively to the payout, and therefore reducing communication and forcing the United Kingdom to develop its own R&D initiatives.

The proximity of Canada to the United States makes each visit return double in terms of information and education; without that proximity, the trips to the United States would probably be reduced. Similarly, visits from Australian, New Zealand, African, and Far Eastern personnel of J&J are more frequent in the United Kingdom because of the ease of passing through that country while on visits to the United States and Canada. On the other hand, U.K. personnel do not readily go to Australia, and African and Indian personnel tend to be trained in the United Kingdom.

The importance of visits is exemplified by the nonwoven fabrics unit in the United Kingdom's having a man at the Chicopee facility frequently, and that one from there is in the United Kingdom every two to three months for an exchange of information on projects. Additional visits occur as officials pass through to professional meetings in Europe. In baby products, visits from the United Kingdom to the United States are more frequent than from the United States to Portsmouth. Lab officials observed that phone calls among professionals who had not yet met personally were more difficult to use successfully than if the individuals had met; appropriate information was not easily obtained without personal acquaintance. Consequently, they argue that more visits should be made between New Brunswick and Portsmouth as a high priority.

The communications sequence begins when a professional scientist, after reading a report, sends a telex to open a channel of communication on a project and obtain more specific information. Letters may be used to obtain an answer to a very specific question, but the telephone advances the time sequence and permits a give and take that is necessary when communication is not wholly satisfactory in written form. A letter is restricting in content and in time. In addition, some questions can be asked orally that simply would not be asked if put in written form.

Although there are no formal or extensive joint projects between the U.K. and U.S. labs, there is an informal one on shampoos, with each independently looking for similar products to meet similar needs. Collaboration occurs where each sees that the other has done something of interest to it or has developed an expertise that it can employ. J&J lab officials do not foresee the likelihood of major projects being cut into separate pieces in the consumer area, for the markets are too different. They are sure that joint projects could be made to work, but they simply are not applicable. In addition, there are some difficult problems of timing, resource allocation within the labs, and goal identification. Further, each is likely to pursue different engineering and manufacturing processes and material routes—for example, pulp technicians in the nonwoven fabrics area see problems quite differently. Finally, the lack of a central research unit in J&J has meant that there is really no place to set up a joint project and run it with a single coordinator. Each R&D unit sees itself as initiating projects in an important, though necessarily competitive, sense.

A type of coordination exists to another lab and the discovery of work that may be already underway there. Even with such subcontracting, no funds may flow, but added experiments are simply undertaken by one lab for the other.

In the toxicology and safety areas, joint projects are undertaken on an informal basis. Still, since the local marketing divisions finance the R&D programs, they do not eagerly approve the use of funds to support U.S. market objectives—though of course each lab does contribute time and equipment without cost when requested by another. These requests are infrequent, however. Both the U.S. and U.K. labs have retained a degree of detachment and autonomy that induces a feeling of potential partnership, rather than dominance by the parent company lab.

Coordination does exist through the elimination of duplication by the U.K. labs' knowing what the U.S. program is even before their own projects have started. Quarterly and semiannual reports are exchanged with sufficient information to permit each country to inquire more thoroughly into what the other is doing and how far it has proceeded.

Despite repeated questioning, J&J officials could not think of ways to quantify the value of the support and background knowledge they obtained from the parent labs. "J-cloth" has simply been copied from Canada, and other products have been lifted technically "lock, stock, and barrel," while others have been adapted to the U.K. markets. Thus, the adhesive bandage from the United States was not accepted in the United Kingdom because the method of adhesion required reformulation in the United Kingdom. U.S. baby lotion was also unsatisfactory for the U.K. market and required reformulation; the U.K. product is now widely adopted by other J&J affiliates around the world.

Relations with Science Community

J&J officials observed that it was quite necessary to have a scientific community within the host country to provide support not only to a scientific attitude on the part of lab professionals but also to governmental regulatory agencies who needed to have impartial advice and counsel. This community would also be used to obtain consultants not only for their own expertise but to add to the reputation of the company in its negotiations with these same regulatory agencies. The J&J labs spend £22,000 per year for consultants. Some science experts are retained permanently, and some medical experts are retained for regulatory advice and to screen advertising. They come from universities, hospitals, and private institutes.

The flow of communication with academic scientists is more from them to the J&J labs than in the other direction. Although J&J professionals do give seminars or lectures at universities and do publish some of the results, neither is frequent. A company secrecy agreement reduces the eagerness to publish, though permission to lecture to professional groups in universities is more readily given. The company's contribution to the British scientific and technology community is therefore direct, in the form of employment, rather than indirect through broader communications.

One of the most helpful aspects of the scientific infrastructure in Britain is the National Lending Library Service, which draws from a variety of professional libraries throughout the country and provides a quick source of scientific and technological information. At times the Library Service draws on J&J resources, as well as those of other companies. Without this resource, the information budget of the company would be substantially larger.

CANADA

Although the British company was the first overseas affiliate, J&J's international growth began in 1919 with the purchase of a company in Canada. It was, until the early 1950s, simply a manufacturing extension of U.S. operations, copying in detail but with considerable lag what was done in the United States. Two engineers were responsible for making certain that U.S. products were precisely duplicated in Canada. No scientific effort was mounted to organize the transfers of technology—much less of R&D results—and the small quality assurance unit did not thoroughly understand the scientific basis of the products manufactured. Now the Canadian affiliate has a full-scale development lab that does some applied research in new products.

Origin

In 1953, General Johnson, president of the U.S. company, ordered the establishment of a lab in Canada. The Canadian management, which did not see the need to expend funds for such purposes, had procrastinated over its establishment. General Johnson saw the need to use local raw materials and to adapt them for J&J production in Canada; his oft-stated reason was that "not all the brains are in the United States."

The first director of the lab hired several Ph.D.'s to generate new products, but their mission was altered fortuitously. First, they did not know enough about the products of the company to be able to generate new ideas, so they had to go into the plant to find out what it was making and how it was done. Secondly, they had to locate in the plant simply because there were no facilities for a separate lab. Once in the plant, they found that they could help improve production and reduce costs in several ways. Quality assurance and technical service was brought under the lab director, removing the conflict that arises from having it report to the manufacturing manager.

The first task of the new lab was an inventory of plant products and procedures and marketing approaches. It required the lab three to four years just to find out what standards were being used in production and the uses of the products in the market. These first years were spent in writing up specifications of what was done in the plant and then helping to change procedures to accord with the higher U.S. standards. Several trips were required to the United States to learn the fundamentals of quality assurance, testing techniques, and what test equipment was needed. These trips led to further efforts by research managers in the United States to persuade top management in Canada of the necessity to support the research lab. (Only sales managers had been promoted to top management in the Canadian company, and they were not technically oriented.) Management had to be sold on the desirability of strict quality assurance.

The job of writing manufacturing specifications and standards was, of course, necessary for the pursuit of quality control, not just for the J&J plant but also for suppliers. Suppliers frequently shipped poor materials, but they could not be rejected because no standards had been set. Manufacturing managers had to be sold on the potential cost reductions and sales increases from the pursuit of standards. For example, the company produced four or five different absorbent cottons without setting different production standards or uses.

These early years were full of frustration for both top management and the new R&D director. The latter coult not get started on new product development because the lab spent so much time catching up on past problems in the manufacturing plant. (The early head of the QA unit, being a chemist, was also the plant nurse, since he could produce mercurochrome and iodine.)

As is frequently the case, the first R&D director hired the wrong kinds of personnel; he wanted creative researchers, but he needed industrial engineers and technicians. Some of the scientists became rather frustrated. Manufacturing was frustrated on its part because QA officials rejected goods. In one instance, top management was persuaded to burn an entire batch of adhesive bandages that were below standard; this act shook up the Manufacturing Division so thoroughly that no such errors were repeated. The problem in the plant was that manufacturing personnel were not attuned to what the QA officials were saying about production processes and product standards. This required that the R&D personnel be continuously in the plant; they were thus prevented from doing research. Several years passed before Manufacturing hired personnel with sufficiently high training so that they could understand what quality assurance personnel were saying. The lab invited the big names in the quality assurance field to speak two or three times a year to the plant managers on the issues involved. Consequently, R&D personnel had to spend an inordinate amount of time during these first years in quality testing, which was still quite rudimentary in the plant. (As several lab officials indicated, quality assurance practices even in the United States were virtually nonexistent before World War II, and have obtained their major support only within the past twenty-five years.) The QA director attended meetings of U.S. officials at least twice a year, getting unlimited assistance from his U.S. counterparts. He would buy test equipment similar to what was used in the United States if he could afford it, and if he could not he would have to adapt specifications or send samples to the U.S. labs. This copying of U.S. practice for the first fifteen years was so close that the lab even perpetuated the mistakes made in the United States, not knowing any better.

Presently, however, the present head of the microbiology lab is giving seminars at various Canadian universities to show the role of quality assurance in industry. In addition, the company has applied standards to purchases of ingredients, setting up a list of approved suppliers. Many Canadian companies have complained bitterly that the specifications are too harsh, but they have gradually come around and are now pleased that they are able to supply other companies with higher-quality inputs. (In one case, J&J forced a supplier to sign a contract with a technical research consultant to improve his products.) This constant improvement in quality assurance standards is itself possible only because of the existance of an R&D lab to back it up. Since QA has to set standards, run tests, evaluate the tests, reject product runs, and train manufacturing personnel, they need a strong research backup.

Successful QA work, the lab managers observe, is not a matter of confrontation but of help; so it quickly moves into technical service, and from there product improvement and modifications naturally grow. (Growth in the early QA-TS stages of the lab can track closely with the rise in production, since a single technician can be hired to do additional work; as the lab responsibilities move into product development, however, teams of researchers must be hired, making the growth of the lab come in spurts.)

In J&J-Canada terminology, operations development relates to the daily activities in the plant plus the scale-up of new processes. Some 50 percent of the time of personnel in these units is spent in the plant itself. Product development concerns changes in existing products and processes; technology development is applied research, related to new techniques and products. Technology development is not needed in the health care lab because adhesive research is better in the United States, and the Canadian lab can draw directly from it. Too, the Canadian market is not significantly different in this product line, and further, the director of this unit is himself not yet as capable in the research field as he is in process engineering. Product development in the health care lab is oriented to new business opportunities out of the J&J line.

The lab began new-product translation only in 1958-1959. R&D personnel made frequent visits to the United States, bringing back sufficient information for the manufacturing plant to begin to recognize the value of the lab's contributions. From this time on, the lab could begin to turn its attention to product developments. To help make this transition, the lab brought U.S. R&D managers and company managers to Canada to give seminars and to meet Canadian management to persuade them to support the lab's work. One of the tactics to gain management's attention was for the QA unit to be more strict than it needed to be in approving products, forcing management's concern for proper manufacturing processes.

It took ten years for the production personnel to realize that they themselves were responsible for quality assurance rather than pointing to R&D personnel when any errors occurred. In order to get the cooperation of manufacturing personnel, some of the R&D staff had to resort to a variety of administrative tricks. For example, the director of the lab never wrote a report on a QA error that put the blame squarely on manufacturing personnel; rather he would cite spurious technical difficulties that had created the error. He would then orally explain the problem to Manufacturing, and they would not repeat it. Manufacturing remained reluctant for some seven or eight years to spend much money on quality control and assurance, but finally, in 1960, they hired professional people for the plant, releasing the lab to do its development work.

In sum, the lab could begin a technical service program oriented to cost reduction only after three or four years of strict quality assurance work, and it was only after another five or six years that it could move to product adaptation. After fifteen years of growth, it was ready to adopt the U.S. program of "Good Manufacturing Practice."

Organization

Besides quality assurance, the lab also has a health care lab and a fibrous products lab. The fibrous products lab has four units concerned with technology development, two with product development, and one with operations development. The health care lab has three on product development and one on operations development, with another responsible for evaluation of competing products.

This organization reflects the transitional situation in the Canadian lab. Having taken care technically of the present manufacturing operations, greater emphasis can now be placed on product development and entry into new market areas, under some long-range planning requested by the president of the company. The lab will now begin to pay more attention to its relationships with Marketing, which is asking for new products from R&D, while the R&D lab is asking Marketing for indications as to what the market needs. The organizational structure of the lab has had to change with its mission and relationship to the rest of the company. J&J officials stressed that R&D personnel are not born, they have to be made. Not only must they be made in the sense of learning how to do industrial research, but they must be fitted into an existing organization that has its own goals and procedures.

The organization of the lab has to be consistent with the goals of Manufacturing and Marketing. As R&D begins to look at possibilities for new products, and even the invention of new products, a corresponding shift towards innovation must occur in both manufacturing and marketing divisions. The close tie between the R&D lab and the marketing divisions is illustrated by the vice president for research's also being the manager of industrial marketing, one of the four major divisions of the company—which also include consumer, hospital, and dental products. In addition, he brought marketing experience gained from two years as marketing manager for one of the divisions.

Lab Capabilities

The capabilities of the lab are illustrated by the background of the personnel, the lab facilities, and the scope of its activities.

The Canadian lab recruits its personnel from among several universities, largely in the province of Quebec. Recruiting is irregular, since it is not customary, as in the United States, for companies to recruit at regular times dictated by university schedules. The orientation of most university-trained scientists is toward academic-type work, however, since there usually is little industry orientation in university programs.

Another source of recruitment for the lab is from other labs in industry—about half of which are labs of U.S. affiliates in Canada—who are closing their labs as a result of poor R&D management, the lack of education oriented to industrial research that provides untrained or academically oriented personnel, and a poor appreciation by company management of the role of R&D. (Lab officials estimated that only about 200 Canadian companies or institutes have research labs of more than five persons, excluding university labs.) Labs are closing because of the lack of company support that results from their not being productive in respect to market needs.

In order to provide a better orientation of personnel, a six-month apprenticeship is provided for each university graduate coming into the lab; this apprenticeship is as a technician, working with a bench scientist. The trainee is then required to write a report on business opportunities in the specific product line to which he will be assigned. In addition, seminars are provided by the marketing & manufacturing divisions on their work and relationship to R&D. To provide a management orientation to the scientists, there are some in-house programs on management by objectives, and there are others on communication and planning skills. If any of the employees wish to go further with their formal degrees, the company will support university work by paying tuition and costs upon successful completion of the courses.

The research capabilities of the lab are supported by a library. A rather low expenditure is permissible here since the company has the resources of six or seven nearby universities, as well as some public libraries and those of the National Research Council and the Pulp and Paper Research Institute. The National Research Council will provide abstracts or copies of articles on request, and has a data bank on a number of scientific disciplines. Some eighty or ninety journals are kept in the library in such areas as chemistry, industrial news, general science, textiles, environmental science, foods and drugs, materials engineering, packaging, medicine, and plastics.

The lab's budget has been increased by over 80 percent during the years 1972 to 1976, increasing total personnel from 80 to nearly 100, of which half have been in R&D activities (with the rest in quality assurance). The lab cannot assimilate more staff than it has now, though it has been pressed by the President to expand even more rapidly. The marketing divisions have also tried to get R&D to expand its marketing research activities to reduce risk in the adoption of new products. They have also tried to dictate the specific R&D projects, down to how to test them in the market, forcing the lab to finally obtain a clarification of responsibilities and authority. It was able to do so partly because the budget of the lab is not allocated to the marketing divisions.

The total program of the fibrous products and health care labs (excluding quality assurance) is roughly divided one-third each among operations development, product development, and technology development (applied research). Thus operations development, which is mostly in the plants, relates to the processes for adhesives, nonwoven fabrics, and toiletries and baby products. The technology development areas are looking at new items and new materials; as they are discovered, they are picked up by product development units for innovation and commercialization in the areas of hospital absorbents, nonwoven fabrics, toiletries, feminine hygiene, surgical dressings, air filters and adhesives. Each of the two labs has its own pilot project that permits a scaling-up of lab work. Health care products are produced by the batch process and are easily scaled up; but fibrous products require continuous production and a separate pilot unit is required, which is considerably more costly. There are no pilot plant facilities for toiletries, and it is much more difficult to make changes in these products—as with adhesives, which are picked up unchanged from the United States. To do any work on adhesives would require a separate polymer group, which is costly and would not be as productive as that in the United States.

However, because of the broad range of nonwoven fabric technology, the U.S. companies have not covered the full potential in research; the Canadian lab has found a speciality into which it can fit. Its program is facilitated in that three segments of activity—the (Chicopee) producer and the two users (J&J in Canada and Personal Products)—are all parts of the same company.

The total number of personnel in these two labs is forty-nine, plus fifty-one in quality assurance. Twenty-three of the forty-nine in research and development are professionals, with the remainder being technicians and other support personnel. Of the twenty-three professionals, four have Ph.D.'s and work mostly in the technology development units. The disciplines represented include chemical engineering, physical chemistry, polymer chemistry, analytical chemistry, microbiology, and mechanical engineering.

Although the lab is not comparable in any sense with those in the U.S., because the missions are quite different, the excellence of its contribution can be seen by the operations development units' having provided trouble-free manufacturing operations and a major contribution to cost reduction. In addition, new products have been introduced into the market through J&J operations. Although the lab will continue to grow with the expansion of the company, it has already reached the critical mass necessary to permit it to do applied research. This critical mass is difficult to identify, but it is characterized by the necessity to have at least one specialist for each technology required in the lab, though if the technique is insignificant in company operations, as in dental techniques or metallography, they can lean on U.S. labs. The critical mass is also determined by the narrowness of quality standards, since the more strict they are, the more expertise is needed in the lab to support the QA unit. In addition, the critical mass depends on the rapidity of technological change and new product innovation, plus environmental pressures and government regulation. Each of these will require personnel to be added merely to keep up with the competitive and governmental pressures.

The Canadian company has no policy of encouraging or discouraging publication of R&D results. Any publication of work requires a release from both the technical and law departments in the company, and little of it has been done. Publication by U.S.

labs occurs in order to obtain acceptance by the scientific community and an entree for those who are doing basic research in the field. This is not as necessary in Canada, and therefore the excellence of the lab is not tested or illustrated in this way.

Projects

There is a company-wide commitment to *use* R&D results rather than merely to pursue research for its own sake. Clear objectives are set, ranging from the development of a new product idea down to the substitution of a local input or ingredient. The distribution of development activities as indicated above is roughly one-third to each of operations development (OD), technology development (TD), and product (new and improved) development (PD), with a gradual increase in the time allocated to new products.

Although equal in budget cost, OD projects number 193, TD projects 78, and PD projects 20, of which two are quite large. The PD category is by far the most costly per project, while OD consists of smaller, more numerous, projects related to manufacturing operations. TD, like applied research, seeks new avenues of activity for the company.

The nature of the projects undertaken by the lab is largely determined by the interests of the marketing divisions. A small percentage of the total effort (around 10 percent) is still given to new product development, though this is gradually increasing. The largest commitment of resources is to the Consumer Division—65 percent of all activities, excluding technology development projects; yet within this 65 percent only 2 percent is devoted to new products, with over half to the improvement of existing products. The R&D lab must show the Marketing Division the place of any given product in a market, and then marketing may be willing to pull that product into its plan.

In the nonwoven fabrics area, J&J sought a "poor man's" technique for producing a product useful to the limited Canadian market. The U.S. company had produced a fabric by methods too expensive for Canada, and the latter needed a product that could be produced on existing equipment and cost no more than $250,000 to tool up. The project was defined also with the requirement that the new product should fit at least two product uses in the existing J&J line to avoid developing a completely new market. Through the ingenuity of some of its researchers, it was able to produce a new product through an air-blowing technique, experimentally employing a redesigned vacuum cleaner for lab equipment.

Projects for the lab are selected by the process illustrated in Figure C1.2. The director and two senior managers of the labs discuss the projects' objectives with reference to total business and the market, and in comparison with what the United States is doing. Idea generation begins with top R&D management brainstorming, which defines a strategic objective, for example in nonwoven; then the segment of the nonwoven business to attack is

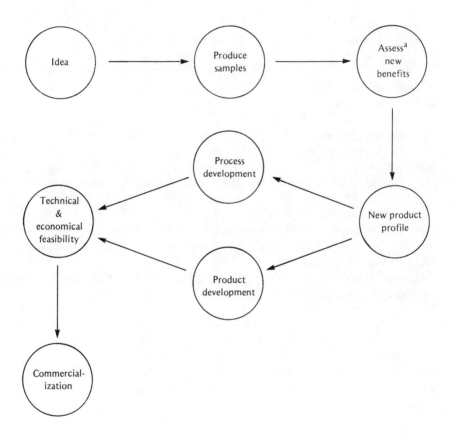

[a]Via:
   —Lab data
   —In-use tests
   —Panel tests
In three areas of current significant business with widely different needs

Figure C1.2. Product/Process Evolution in J&J's Canadian operation.

pinpointed, along with which U.S. product to take off from. Some of these decisions are stimulated by the expiration of patents on existing products or processes, meaning new developments are needed to protect market position.

The gestation of a TD project illustrates the steps required: In the generation of a new idea, the first check for approval will be with the president of the Canadian company, to explain the market-orientation of the idea and the absence of duplication with the parent company in the United States or other affiliates of Johnson & Johnson. Assurance will have to be given the president of a con-sumer need, obtained from investigations by the lab itself. Although meetings are held with the marketing manager in the product area involved, it is not likely that at this early stage a positive response will be obtained, since marketing managers may not be readily persuaded as to what can be done with *new* products. Therefore, R&D has to go all the way through a feasibility project exercise, producing a prototype and doing consumer testing when marketing is not responsive. In PD projects, the lab goes directly to marketing, since much of the activity is on improvement of existing products.

Once the lab has completed the prototype and market tests, it is responsible for the cost of the first couple of production runs or batches, recouping its cost only when the test trials are successful, and the product can be sold. The lab actually rents the production facility for these two runs, paying all the factory costs, including the cleanup of the facilities.

Finally, the responsibilities of the lab extend to regulatory affairs, taking some 10 percent or less of the time in each area. Even the quality assurance section spends some 10 percent of its time in regu-latory affairs. The time of this section is allocated some 45 percent to in-process and finished goods testing, 25 percent to raw materials, 10 percent to microbiology, 10 percent to specifications, and 5 per-cent to contract manufacturing, in addition to that for regulatory affairs. As indicated earlier, assurance constitutes about half of the total lab activities.

Coordination with U.S. Labs

Although the Canadian lab is clearly separate from the activities in the United States, and no one in the United States has tried to tell the company how to run its QA or R&D activities, the lab has been able to get anything from the United States that it knew how to ask for. Its officials have been treated as colleagues on visits to the United States, rather than as second-class citizens, partly because of the lack of desire on the part of J&J to force standardization or uniformity among affiliates. In the assessment of Canadian officials, though, the lab simply could not have gotten where it is now without assistance from the United States, because it could not have succeeded on its own or have found similar assistance within Canada, for such knowledge did not exist.

The lab maintains communication with other affiliates around the world, with the quarterly reports being sent to other labs; any inquiry from abroad is answered promptly by letter (the report of the professional at the bench would not necessarily be sent unless it happened to fit precisely with the needs abroad). In addition to the exchange of reports, four joint committees cover the fields of absorbents, fiber engineering, fiber research, and regulatory affairs. The director of the Canadian lab sits on the committee on absorbent technology in the United States, and a British counterpart would be included if distance did not preclude it.

Personal visits are extensive and are urged by the Director under his program of management by objectives. Each project leader and senior manager must take at least one trip to the United States each year and talk to two or three of their U.S. counterparts. They do not generally go to Britain. However, British, South African, Brazilian, and other lab officials detour by way of Canada on U.S. visits. These visits from other affiliates total between forty and fifty a year (some 260 foreign officials visit the company as a whole during any year). Such foreign visits eat into a work schedule, knocking it into a cocked hat at times, since visitors are not always scheduled very far in advance and often stay for two or three days. Visits from U.S. lab officials total between seventy and eighty during a year, from project leaders up to top managers. In one month, the Director himself had to receive six visitors looking at R&D management. This is as it should be, in his view, for such visits are most productive both ways. These discussions at top management level include not only getting a line on what is happening but also on problems of managing personnel, relations with marketing, and possibilities of dovetailing research programs.

One joint project arose out of pure happenstance. The lab had been working on nonwovens with an air filter equipment company and had to buy some new equipment that seemed to be useful in other materials as well. Some of the researchers got curious and tried a number of alternatives, eventually getting pushed into the aerodynamics of the problem, which a new researcher was interested in. Out of many experiments came a new machine for manufacture of unique nonwovens (for diapers and other products), but the lab failed to persuade Manufacturing that it would be a useful new technique for Canada. It was, however, picked up by J&J-Chicago, one of whose engineers spent six months in Canada working on the project, taking all data back to Chicago.

There is no dovetailing of PD or OD projects with U.S. labs and no subcontracting with any of the affiliates in these activities. In TD projects, once Canada has advanced in a particular idea, the United States might look to Canada for continuing results, not pursuing that particular route itself. Duplication is not a critical problem, since each is aware of what the other is doing. The more serious problem is that things developed by the United States or Canada are not picked up by the other when they would be appropriate.

Still, the Canadian lab relies heavily on the United States in product development, and adaptations are made only when necessary. An example of an adaptation is a new hygiene product that will be a *second* generation of an existing U.S. product; it probably will be picked up in the United States once it is completed in Canada. One of the reasons for the lack of modification is that clinical tests on adaptations are quite expensive, so they will not be done unless absolutely necessary. The lab had to learn the hard way to stop trying to reinvent the wheel and to accept what was done by others. It finally defined its purpose as that of attempting to "enact business" and not to "invent," unless necessary.

The nature and extent of coordination and assistance with the United States varies considerably according to the stage of the lab's development and the kinds of activities it is engaged in. For example, there is presently little exchange in the operations development (OD) area, but this was not the case earlier. It takes time for a new lab to determine its mission in relationship to other labs and to become sophisticated enough to be able and willing to take advice and help without feeling that it is a second-class citizen. This growth requires personnel who are themselves adaptable to new situations and changes.

The ties to the U.S. operations in the OD area (which is similar to technical services) are with production and manufacturing operations, from whom information is sought on the least costly processes and quality maintenance. Specifications emanate from the United States, and quality standards are high. In the lab's first stages, when most of the work was on operations development, it needed considerable support through direct documentation on processes and engineering from the U.S. labs. Even now, some changes in operations would be much slower if the lab were cut off from the United States.

Damage to the product development (PD) program would also occur if the lab was cut off, since it is picking up a number of prodcuts coming from the United States. However, it is beginning to initiate some of their own activities independently, and would be able to move along this line if separated. PD projects are tied closely to Marketing, and the project leader must know all attributes of his product including technical, market, and consumer aspects. (PD projects have to be approved by the marketing director if they cost more than $2,000; under $2,000 or six months' time, whichever comes first, R&D can do any feasibility project it wishes.) The project leader is in constant contact with his opposite number in the United States and in other affiliates through trips and by phone—two trips a year at a minimum, which the lab would like to increase to two per quarter. These visits are most profitable because of the ease of obtaining information from J&J labs in the United States. (One of the senior managers indicated that the U.S. facility was considerably greater than the R&D labs of the company with which he had previously worked in Canada.)

The lab is also able to take initiatives in the TD area, but it relies on substantial inputs from the United States. The flow of information is open and important, and the U.S. managers make the Canadian officials feel like equals, making them more interested in the exchanges. There is some danger to the development of the lab in being so close and potentially dependent on the work in the United States, but this disadvantage is offset by having full knowledge of what is happening in the States and being able to work from it. TD projects are less tied to the marketing divisions and much more closely related to U.S. activities in R&D, especially through the joint committees mentioned earlier.

One of the complications of the closeness between the U.S. and Canadian operations is that the marketing divisions in Canada are themselves very strongly tied to marketing divisions in the United States—more closely than to the Canadian R&D lab—taking signals from the United States as to what they will do and what to accept from the R&D labs. Marketing tends to reduce its risks by following the U.S. pattern rather than taking initiatives from the Canadian lab.

One final area of cooperation between the U.S. and Canadian operations is in the patent program. Assistance from the U.S. Legal Department is given to Canada by two lawyers assigned to Canadian projects, who are brought in at the earliest stages of an idea. This early participation assures their interest throughout the project in the determination of the patentability of the results. The lab could not find this kind of assistance anywhere in Canada simply because others do not know the business in the same way. The U.S. Legal Department also does the patent searches and helps in the formation of the patent application.

Contributions of the Lab

The Canadian lab has made a number of significant adaptations to Canadian needs and has developed a few indigenous products. Lab officials asserted that they would like to be judged on the following contributions:

> The number of new products brought out for marketing, as well as improvements in existing products;
> Patents per thousand man-hours of work in PD and TD projects;
> Contribution to profitability in cost reduction or in defending against competition;
> Protection of the business through regulatory affairs—both through meeting regulations and helping to set regulations.

In terms of regulatory affairs, the lab helped set up the dialogue between the Canadian government and the Medical Manufacturers Association; the director of the lab spends some 20 percent of his time in this activity that, when added to that of others in the lab, amounts to about two man-years. Much of the representation before regulatory bodies is based on U.S. documentation. In 1976 the lab was represented on the Government Standards Committee examining "ethylene oxide sterilization"; along with other members representing U.S. affiliates in Canada, all were relying on U.S. documentation. In addition, the microbiologist from the lab is on the steam-sterilization committee in the government. The next steps in regulatory affairs will be to develop clinical research capabilities within Canada. The lab is not experienced in this, yet the capability will be necessary in the future in order to play a more effective role with governmental agencies. There is an increasing Canadian government concern to have clinical research done in Canada rather than relying on that in the United States.

The contribution of the lab to the company is seen also in diverse inputs with Purchasing, Marketing, and Manufacturing in the setting and pursuit of company objectives. In addition, complaints from consumers have reduced considerably, and rejection of particular batches or product runs have also been significantly reduced. Government purchases from the company have increased significantly, and not one rejection has yet occurred by the Canadian Consumer Protection Agency.

From the standpoint of the national economy, a number of high-quality local materials have been successfully substituted for imported products, despite frequent somewhat higher quality of the imports.

The development of nonwoven fabrics for tea bags is an illustration of the lab's success both from the standpoint of the company and of the economy. The lab had tried to heat-seal gauze for tea manufacturers, and it dawned upon them to try nonwoven fabrics. Gauze had been used because it was without the taste, that paper bags produced in the tea. But gauze had become too expensive; being able to heat-seal the nonwoven fabric similarly to paper provided a 50 percent reduction in the cost compared to gauze. One of the major problems that was solved was to eliminate all taste coming from the nonwoven fabric that would alter the taste of the tea.

Future

The lab will expand its work in technology development (TD) and in regulatory affairs, with the rest of its activities paralleling the development of the company itself. To meet its new responsibilities, a new building and facilities are being added, doubling annual capital and equipment expenditures for a few years, but adding only five to ten percent to operating costs per year. The new facility will provide a place for pilot operations, which will be scaled down to one-half or one-third the size of unit-manufacturing facilities in the plant. (For example, one set of equipment for a batch process or production line will be half or a third that of the plant capacity for a similar unit.) This improvement in pilot facilities is necessary to carry the invention through the final processes, and thereby persuade Marketing and Manufacturing of its usefulness and feasibility.

The mission of the lab for the next several years has been characterized by its Director as follows:

To ensure that consumer needs and benefits are identified as a basis for all product innovations.

To provide the technical resources for new products as required by corporate growth objectives.

To engage in market-oriented research only and not carry out contract research or duplicate projects that meet our needs and are being actively pursued by affiliates.

To ensure, by Quality Assurance's use of meaningful specifications and quality procedures, that current products will continue to meet our quality standards.

To do no work other than adaptive and supportive work in the area of dental products.

To monitor all regulatory and compliance affairs related to products.

To carry out technical, long-range forecasting and related planning.

To service the ongoing technical demands of the operating divisions.

Relations to Scientific Community

The lab has not been able to rely significantly on the Canadian science community, simply because there is little academic interest in industrial research. Nor has the lab been able to ask suppliers to do research related to materials, though the suppliers have at times had to modify their products, and have had to do some research on their own in order to know how to comply with J&J standards. Suppliers often rely on their own U.S. parent companies for such adaptations. A wider science community in Canada would take some of the load off the R&D lab.

The Canadian lab does have a few consultants, but only two on retainer—one in clinical research and another on a specific research project. Others are retained on an ad hoc basis.

The Canadian government itself has given virtually no support to the development of the lab. Over its twenty-two years of existence, it has received only $35,000 in support grants. It has been virtually cut off from such assistance because of its ties with the U.S. company, which prevented exports from Canada into U.S. and European markets. But the company in Canada has not sought to export only because its transport costs are too high.

BRAZIL

Johnson & Johnson do Brasil has sustained a tenfold growth over the past decade, as a result of both product line expansion and the growth of the market. Some 1,000 products are made by the company, including 400 in the Ethicon Suture Division.

This growth brought a divisionalization of the company, including R&D activities, with each Division sponsoring what it could afford to support. The Consumer Division produces some sixty products, including baby oil and shampoo, a baby cologne, talcum powder, a baby soap, diapers, safety pins, disposable diapers, toothbrushes, a skin-cleaning lotion, an absorbent cleaning cloth, and a variety of products for feminine hygiene.

The company was begun in 1943, but it was only after World War II that quality control labs were instituted, with the added responsibility to adapt any new products added to the line.

In 1955, a Director of Research was appointed for the company, reporting to the General Manager. He began a wider effort at adapting a variety of products, beginning with adhesives, bandages, and baby products. Quality control units were also placed under his responsibility. Even so, by 1960 there were no more than thirty people in the various technical activities. In the early 1950s, pharmaceuticals were added to the line, and later some local pharmaceutical products were developed. By 1972, the R&D lab was divided into its present sections, and a new building was provided to house both of them.

Organization

The company has two major divisions: consumer products and hospital, and medical and diagnostic products (see Figure C1.3). These in turn are subdivided; the first into consumer products (personal and baby products) and health care and hospital; the second into medical-diagnostic; and Ethicon sutures. Each of these subdivisions has its own R&D facilities.

The Brazilian company, thus includes something in virtually all of J&J's product areas. Since this study of R&D activities has focused principally on consumer and health care products, most of the illustrations of Brazil's activities are also drawn from that division. However, since the labs are housed together and do have some joint projects, some of the activities in the medical-diagnostic areas are included.

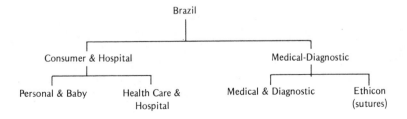

Figure C1.3. Organization of Johnson & Johnson do Brasil.

Virtually all of the R&D activities are housed in a new building on the plant site some 95 kilometers northeast of São Paulo at San José dos Campos. R&D in industrial products is in another building, and some pharmaceutical research is in a third location; a chemicals lab is at Sumare, where chemists (some Ph.D.'s.) are working on improving the synethesis of pharmaceuticals. The main lab holds the research activities for the Hospital and Health Care Division (and the director of this lab is responsible for the administration of the building), plus the R&D lab for Consumer Products, and quality control units for both. It also contains the pilot plants for baby products and consumer products, biological research and quality control, QA for pharmaceuticals, the training center for blood banking, serum research, veterinary research, the catheter plant, a diagnostics production line, and computer facilities.

The lab for health care and hospital products has five divisions: adhesives and plastics, plus packaging; hospital products and fibers; health care and toiletries, quality control; and building administration.

The consumer products R&D is divided among units concerned with a variety of products, including absorbent products, and toiletries. Each department, as well as quality control, reports to the managers.

By comparison, the R&D activities in the medical-diagnostics division include quality control, selection of raw materials (human serum and blood cells), physical control, chemical analysis, biological control for the entire company, and administrative support activities. The biological control unit supports all of the labs in this function and also has responsibility for water and effluent control.

Each of the labs reports to the head of the subdivision to which it is attached, making quality control separate from Manufacturing or Marketing. At the same time, the responsibility for each is tied directly into the division managers' own responsibilities for profit performance, giving a commercial orientation to the activities of the lab.

Facilities and Staffing

The major lab building has over 20,000 square feet of space, half of which is dedicated to quality control activities. The lab's R&D facilities include a classical analytical lab and modern instrumentation, e.g., infrared spectrophotometry, gas-liquid-chromotography, spectrodensatrometry, ultraviolet spectrophotometry, and the like.

To provide the literature searches needed in research and testing, the lab has a library stocked with thirty different journals in the chemical, medical, engineering, cosmetology, plastics, and hematology fields—80 percent of which are from the United States, with the rest from Germany, Britain, and Brazil. The library is further supported by some 200 Qualified Libraries throughout Brazil, which are cross-indexed through a BIREME system in São Paulo; each of the member libraries will supply an article or abstract within a week at a nominal cost of Cr. $15. The lab usually finds what it wants in Brazil through one of these labs; if not, it can obtain it within three weeks from one of the J&J laboratories in the United States. The library is tied to a data bank in Maryland called "Medline" for abstracts on any topic. It can obtain this information through satellite communications, thanks to its close relationship with the engineering institute at San José, thereby obtaining it free.

Qualifications of the staff in the R&D labs has shifted lately from a majority of people trained as technicians to a majority with bachelors' degrees from the various universities—half from the São Paulo State University and half from elsewhere in Brazil. Few of the employees in quality control units have more than technician training. There were no Ph.D.'s at the main lab in San José dos Campos.

The company cooperates with the Institute of Engineering in San José, which provides training up to the Masters level. Three of the professionals were attending courses for eight hours a week at the time of the interviews. An agreement with two universities in West Germany will lead to Ph.D. degrees for some of the professionals: one researcher was there during 1976; she was not previously with J&J, but receives partial support from the company for her study of tropical diseases (Chagas' and schistosomiasis). On her return it is expected that she will work at J&J and teach at the São Paulo State University.

Some 70 percent of the personnel in the R&D areas (excluding quality control) are university graduates. Since it is most desirable to keep them learning, training courses are provided in the areas of microbiology, quality control, and serology; the teachers in these courses are lab managers. Such in-house training is also desirable because it is difficult to get graduates from the best universities out to the location near San José. Professionals do not want to leave the environs of São Paulo, causing a shortage of well-prepared graduates available to the lab. Several of the professionals at the lab teach at some of the universities and are able to recruit from these classes.

The factor of distance is a serious handicap, since many of the professionals are simply not moveable. Though they have some consultants at universities, the lab professionals have to go to see them rather than vice versa. It is also difficult to obtain professionals with any experience in industrial research; it required nine months to recruit one professional with three years of industrial experience. Industrial research needs are mushrooming in Brazil, especially in centers such as San José. But any place outside of metropolitan areas is seen as a poor location, although San José is actually in beautiful country, not too far from the beaches, and will probably be seen as quite desirable when people begin to flee urban blight.

Another way in which the lab is developing associations that may be helpful in recruiting professionals is through training of hospital technicians in the latest technique of hematology and blood banking. It has developed mobile equipment that it can take to the hospital for demonstration and teaching, or the technicians can come to the lab for a program of several days. These contacts are useful not only in a client relationship but also in demonstrating the attractions of industrial research.

The company will be paying increasing attention to the training of its professionals in management techniques as the labs expand in number and responsibility. The present directors of the two labs in professional and consumer products both attended an outside course in R&D management at Slough, England.

The budget for the lab is divided according to quality control and R&D projects. The quality control expenses are based on a market forecast of sales and are charged to manufacturing as a "cost of goods sold." They must be approved by the Manufacturing Division, and usually amount to around 2 percent of sales. R&D projects are presented to the General Manager and the Executive Committee for approval, and the annual expenditures in the professional health care and consumer products divisions amount to roughly one percent of sales, compared with 4 percent at the Domestic Operating Company (DOC) in the United States. However, the labs obtain much assistance from the DOC and other U.S. companies, and provide assistance to activities of J&J affiliates in Argentina, Mexico, and other countries.

Lab Capabilities

In attempting to describe the capabilities of the lab, officials observed that these facilities were J&J's most advanced in Latin America, and will probably remain the center of company R&D activities for that region. The lab is, in their view, one of the best in the country in any industrial sector, comparing favorably with these of Nestlé, Ciba-Geigy, Pfizer, and others. It was probably the first to move into R&D projects, and is one of the leaders in this activity, though others may be equally good in quality control. This strong R&D.activity reflects the company's position as the largest in total sales within its industrial sector, and the fifth in pharmaceuticals.

The capabilities of the lab are attested to by the government's occasional use of the lab as a "reference lab" for the setting of standards and for determining the necessity to recall products of other companies (J&J has not had to recall any of its own).

This excellence is further illustrated by the obtaining of a contract with Brasilia as a result of very careful presentation by one of the lab's scientist on how the operation was set up and production was controlled. In addition, the lab has been used for training of public health technicians on many occasions, for several months at a time.

The programs of the three labs include not only quality control, but also technical service concerning manufacturing processes, materials handling, and cost reduction activities. The second level of activity emphasizes the use of materials inputs, materials handling, and quality testing. A third stage involves mostly the adaptation of U.S. products to Brazilian conditions, materials, and equipment. Some more advanced projects are directed toward the development of products uniquely suited for the Brazilian market.

The workload in the Medical and Diagnostics Lab is usually 65 percent involved in quality control, 13 percent in adapatation of products from J&J-U.S., and 22 percent in the search for new pharmaceutical and diagnostic products. The program of the Health Care and Hospital Products Lab shows a distribution of time, apart from quality control, as follows: 40 percent on technical services on existing products and cost reduction activities; 30 percent on viability studies of J&J products for adaptation; 20 percent for the development of products determined to be useful in the company line; and 10 percent on specifications for quality control work.

Projects

Since the charter of the Brazilian company is essentially to translate J&J products into Brazil, a major thrust of each project is to determine the fit of a particular item in the Brazilian market. This involves an assessment of the availability of materials and equipment, the capacity of the plant, costs, and the necessary price compared to the need and the demand in the market. For example, the translation of the "Stay-Free" feminine napkin into the Brazilian product "Sempre Livre" started with the receipt of product specifications from the United States, including the necessary equipment specifications. Pilot samples were made from Brazilian materials once the necessary equipment was available, and preproduction samples were produced by R&D engineers and the development engineer at the plant. The adhesives had to be changed to use local ingredients, and experiments made to develop proper adhesiveness. The fiber products R&D unit developed several adhesives for sealing the napkin and for heat-sealing processes. A silicon treatment was added to the sides to prevent leakage and a plastic packaging was tested for convenience marketing. R&D engineers were involved in the project for three to four months, while it was being approved by manufacturing. Until that time R&D was responsible, and underwrote the costs of the first month's production, bearing the expense of any waste products.

Two products exemplify product development by the labs. One is a skin-cleaning product called "Higiapal" and the other was a professionally-designed toothbrush. The skin-cleaning product was found almost by chance; the researchers were working on antiseptics and found a synergy between two compounds that had killed bacteria. This result would have made it a drug in the United States, but not in Brazil, permitting it to be sold over the counter. A second market was found for the item in relieving diaper rash.

In the second instance, the company decided to broaden and upgrade its penetration in the toothbrush market, despite this product's no longer being marketed by J&J in the United States. It selected a panel of dentists, asking them to set up the criteria and then to write the specifications for a toothbrush they would recommend to their patients. With these specifications, they contracted an outside designer and obtained the proper molds. The plastics plant then made tests on the material, while the R&D lab selected the bristles and plastic materials, testing the efficiency of the design, which had to be changed somewhat.

Coordination with U.S. Labs

Since almost all of the work in Brazil is related to translation of U.S. products, there is substantial flow of communication from the U.S. labs to their counterparts in San José. And, since the Brazilian lab is the most advanced center in Latin America, it also sends all the information on products that other Latin American affiliates want to copy. These affiliates also send samples for testing to the Brazilian labs.

Communication with the U.S. labs is facilitated not only by the circulation of semiannual R&D reports, but also by the international J&J meetings on special subjects. The International Technical Operations Services also provide information regarding processes and manufacturing equipment, and the international quality assurance service helps formulate specifications on quality standards.

Although considerable efforts are made to maintain high levels of communication, distance is an obstacle in terms of mail delays of from ten days to a month, and the expenses of personal visits are rising rapidly due to governmental restrictions on foreign exchange. Telephone charges are also fairly expensive at $15 each by direct dial, but it is used at least twice daily to the United States or Europe by the R&D managers. Since letters take so long to arrive, they have diminished as a means of communication, despite the cost of telephone and telex.

The researchers at the U.S. and other labs are quite willing to provide all information necessary to get into production, ranging from specifications, detailed manufacturing processes, and even the "little black book" or the "problem book" showing the difficulties in actual production stages that arose in manufacturing.

Personal visits are seen as exceedingly important, since technology cannot be as effectively transferred by correspondence. The director spent three weeks with Ortho in the United States to agree on a schedule for transfer of a given product, and a professional will remain for two weeks longer to pick up all the details of manufacturing. The same R&D manager frequently has been to the United States, Europe, and Argentina, under no serious budgetary constraint. Three of his professionals have been in the United States for more than a month each for discussions in the fields of pharmacy and dentistry.

Despite the distance, the labs and plants have frequent visits concerning the start-up of new products, quality control, technical services, and the adaptation of local equipment to manufacturing needs. Several of the U.S. professionals will stay for two or three weeks to make certain that processes are moving smoothly.

Contribution of the Lab

One of the contributions of the lab has been to improve the quality of supplier products. J&J has frequently given the lab manufacturing specifications and helped it to develop new products—for example, resins needed by J&J. The parent company will show the Brazilian company a U.S. product and request that an equivalent be supplied. Samples are made that must be approved and schedules determined for manufacturing by the supplier. R&D professionals work with the supplier and his R&D personnel (if any), often working with several alternative suppliers to make certain that a good product can be obtained. J&J may have to persuade the supplier to buy special or additional equipment to meet their specifications and volumes. This new technical expertise is then available for the supplier to use in meeting the needs of other customers.

One of the benefits of the lab is seen in the quick approval of products by governmental agencies; this benefit is not measureable, but it is seen in the high cost that would be incurred if approvals were delayed. The government has recognized the excellence of the J&J labs by accepting the standards established by the company in pharmaceuticals, adhesives, and sutures.

Many of the R&D projects have resulted in a substitution of indigenous materials for imports in such items as "fluffy pulp" adhesives, string, polyethylene film, etc. The reduction in costs through elimination of import duties has saved over Cr.$6 million in consumer products alone.

It is much more difficult to change formulations in pharmaceuticals and drugs to permit the use of local ingredients. Even so, the lab has saved Cr.$700,000 in cost reduction in lab operations alone by changing methods and materials—but nothing has been directly saved in plant costs. Eventually, the lab anticipates substantial cost reduction through the use of indigenous materials in a few pharmaceutical products. In general, however, local materials would be costly, and import substitution would not necessarily reduce cost of production in pharmaceuticals.

In diagnostics, several serums have been developed out of wholly indigenous products, eliminating imports. In baby products, plastic bottles were obtained from domestic suppliers after substantial assistance was provided them by the labs, and substantial savings will result. In pharmaceuticals, orvules and suppositories were shifted to local materials with some product improvement, again eliminating imports. This project required one and a half to two years of work with one- to three-man teams working at various stages.

The company had previously imported 480 liters of finished product from Ortho in the U.S., at a cost of Cr. $4.4 million. In the development of anti-Dserum, a high-titer serum is needed requiring blood from donors to be concentrated to raise the titer, which is done through blood filtration. Through lab projects estimating donor volumes and costs, they were able to show that they could obtain the same volume of serum from Brazilian donors at a cost of Cr. $2.2 million or a savings of U.S. $200,000 from one project alone.

Given the continued price control by the government, stringent efforts are made at cost reduction. This problem and that of import substitutions are so critical that it will be necessary for the lab to examine all possibilities of eliminating imports, a pressure increased by the rise in oil prices. J&J is a substantial importer of actives in its pharmaceuticals and medical products. These will increasingly have to be produced locally in the future, and J&J will assist through its research program in developing such suppliers. Seventy percent of consumer products are made from local ingredients (all pigments and fragrances are now Brazilian), while imports are largely industrial grade chemicals, resins, and binders. Brazilian petrochemistry is not yet fully developed; when it is, these items will be obtained locally.

The five-year projection for the labs shows that they will still stay within the present J&J worldwide product line, but there is a desire to get into uniquely Brazilian products. For example, there are licensees for some non-J&J pharmaceuticals. However, price controls are stifling pharmaceutical development—both research and production—making the industry very conservative in developing new products or taking any new initiatives. A substantial part of the future market seems to be in the supply of generic drugs to the Public Health Agency through public bidding, such as the sale of one-minin multivitamin tablets, which are distributed virtually free.

J&J anticipates some substantial changes in the pharmaceutical part of its business, simply because only a few of the 450 Brazilian pharmaceutical companies really do adequate quality control work. The Brazilian government wants these few companies to assist in developing quality control procedures in other companies in order to have an adequate supply of high quality products. J&J is now providing smaller competitive labs with products that they in turn sell to the government. These small companies cannot be bought by international companies, yet they must be supported. Still, some shake-up is to be anticipated.

Community and Government Relations

There is hardly an adequate scientific community within Brazil to help support the work of J&J or other laboratories. In fact, very few of the locally owned companies do any significant R&D, and only a few of the affiliates of international companies. These is also little communication among R&D managers regarding the kinds of problems they face.

As an alternative, J&J has provided an example for other companies in the development of their quality control activities. The head of the Brazil QC unit created a company "QC circle" composed of manufacturing operators, supervisors, managers, R&D personnel, and QC personnel to exchange information on the problems of quality control. This circle meets after hours to discuss means of cost reduction and increase of efficiency, safety, personnel problems, and so on. This exchange of information enhances job satisfaction and raises productivity. Over forty such circles have been set up, and incentives are offered to each based on performance in improving products or reducing costs—a series of free training courses. This model has been adopted by other companies in Brazil, enhancing their QC activities.

The company can use only a few governmental labs for testing. The Institute of the Brazilian Food and Drug Administration does provide tests on purity and toxicity of materials and products. An aeronautics lab at San José provides metallographic tests. Specialized research work, such as that on long-fiber cellulose, has been subcontracted to the government's Technological Research Institute. These same labs provide opportunities for consultation on special questions, as do professors at some of the universities. Expert consultants are retained in the fields of baby products, personal products, and hospital products, as well as for clinical testing. The labs are also cooperating with several governmental institutes in joint projects on diagnostic reagents in several diseases, and in some clinical testing done by the various governmental institutes.

Conversely, several of the scientists in J&J have professorial appointments at engineering and chemistry departments in local universities and teaching assignments at some of the secondary schools.

The development of close relations with university departments has not always been easy, since some of the professors are not happy with the existence and behavior of affiliates of international companies. However, J&J has worked to establish good relations, extending some research grants and providing scholarships to its own employees for continuing education at the universities. One problem in subcontracting work to a university lab is that the universities do not have adequate accounting procedures; another is that most of the professors at the universities would prefer to help smaller companies.

These attitudes mean that the fundamental research triangle—tying government, universities, and industry together—has not been constructed in Brazil. The University of São Paulo reportedly copied the orientation of the Sorbonne in France toward highly theoretical and abstract studies, and the rest of the universities in Brazil copied São Paulo. Consequently, there is an inadequate orientation to industrial research, which will have to be built up gradually. The absence of this fundamental triangle means that it is difficult to attract Ph.D.'s into industrial research, and that recruitment is going to require a program of continuing development of professional attitudes, including trips to headquarters labs to make professionals feel "part of a larger scientific community." But it is just this integration of scientific abilities and technical orientations that the international company is capable of generating, and J&J is finding it desirable to do so.

Index

A

Absolute centralization, as managerial style, 40, 42-46
Acquisition, as entry mode for research and development activities, 33-35
Agency for International Development (AID), 276
Agriculture research, 226
Animal feed research, 248-249
Automotive components industry, 34
Automotive paints, 151-153, 172, 188
Autonomy, 31-32

B

Brazilian National Research Council, 269, 270
British Royal Society, 269

C

Canadian Research Council, 99
Ceytea, Ltd., 234
Color, discernment of, 153
Colworth lab, 203, 218, 234
Cooperation, as managerial style, 40, 52-53
Critical mass, concept of, 73-75, 76, 77

D

De Bodinat, Henri, 40, 52
Detergent research, 217, 247, 267
Developing countries, research programs in, 277-282
Direct placement, as entry mode for research and development activities, 30-33

About the Authors

Jack N. Behrman is Luther Hodges Distinguished Professor at the University of North Carolina Graduate School of Business Administration. He has held faculty appointments at Davidson College, Washington and Lee University, and the University of Delaware, and visiting professorships at George Washington University and the Harvard Business School. In addition, Dr. Behrman is a frequent member of research panels for the National Academy of Science and the National Academy of Engineering; an advisor to the U.S. Department of State and the U.N. Centre on Transnational Corporations; and Senior Research Advisor to the Fund for Multinational Management Education in New York. From 1961 to 1964, he was Assistant Secretary of Commerce for Domestic and International Business. Dr. Behrman is the author of numerous articles, books, and monographs, including *Some Patterns in the Rise of the Multinational Enterprise* (1969), *National Interests and the Multinational Enterprise* (1970), *U.S. International Business and Governments* (1971), and *The Role of International Companies in Latin American Integration* (1972); he is coauthor of *International Business—Government Communications* (1975) and *Transfers of Manufacturing Technology Within Multinational Enterprises* 1976).

William A. Fischer received a B.S. (in civil engineering) and an M.S. (in industrial management) from Clarkson College and a D.B.A. from George Washington University. His work experience includes both industrial and government positions in the management of high-technology projects. At present, he is an assistant professor at the School of Business Administration at the University of North Carolina at Chapel Hill. Among Dr. Fischer's recent publications are articles on technology transfer, scientific and technical information and the performance of R&D groups, and postwar Japanese technological growth and innovation. He is a member of the editorial advisory board of the *Journal of Technological Transfer.* His research interests include the management of technological change, technology transfer, and corporate technological strategies.